Achieving Competence

in

Science

PAUL S. COHEN

Assistant Principal–Supervision–Science
Franklin Delano Roosevelt High School
Brooklyn, New York

JERRY DEUTSCH

Teacher of Chemistry
Edward R. Murrow High School
Brooklyn, New York

DR. ANTHONY V. SORRENTINO

Director of Computer Services
Formerly Teacher of Earth Science
Monroe–Woodbury Central School District
Central Valley, New York

Achieving Competence in Science

When ordering this book, please specify:
either **R 568 P** *or* ACHIEVING COMPETENCE IN SCIENCE

AMSCO

AMSCO SCHOOL PUBLICATIONS, INC.
315 Hudson Street / New York, N.Y. 10013

ISBN 0-87720-018-1

Printed in the United States of America

8 9 10   99 98 97

About This Book . . .

This book is designed to provide students with a comprehensive review of middle level science and enable them to prepare for standardized state or city competency tests in science.

The book contains ten chapters. Chapters 1–3 are devoted to the life sciences, Chapters 4–6 to the earth sciences, and Chapters 7 and 8 to the physical sciences. Chapter 9 discusses energy sources and issues, and Chapter 10 deals with the interactions of science, technology, and society.

The text presents the major ideas of each topic in a simple, straightforward manner, with a minimum of unnecessary detail. Over 130 illustrations accompany the text, serving to clarify concepts. Each chapter is divided into several sections, each of which is followed by an exercise consisting of questions that reinforce the main points presented in that section.

In addition, special features called "Process Skills" appear at intervals throughout the text. As the name implies, the purpose of these features is to teach students a particular process-oriented skill or skills, such as interpreting graphs and diagrams. Each feature guides students through a process skill and then concludes with several follow-up questions that require students to apply the newly acquired skill on their own.

Throughout the book, vocabulary terms of major importance are printed in **bold italic type**. These terms are defined in the text, and also appear with formal definitions in a glossary at the back of the book. Terms of secondary importance and words that may be unfamiliar to students are printed in *italic type*. These terms do not appear in the glossary, but are listed in the index.

Finally, the book includes a Practice Test consisting of 70 questions. The test focuses on the major concepts, understandings, and process skills of a typical middle level science curriculum. Answers to the exercise questions, Process Skill questions, and the Practice Test are available in a separate Answer Key.

Contents

Chapter 10. **Science, Technology, and Society**

Chapter 1. Living Systems: Organisms

PART I. LIVING THINGS AND THEIR CHARACTERISTICS

Living Things Carry Out Life Processes

All living things, or **organisms**, share certain characteristics that set them apart from nonliving things. In particular, all organisms carry out *life processes*, some of which are listed below in Table 1-1.

Table 1-1. Life Processes and Their Functions

Process	*Function*
Respiration	Releasing energy stored in food
Transport	Moving materials throughout the organism
Ingestion	Taking food into the body
Digestion	Breaking down food into a form usable by cells
Excretion	Eliminating waste materials produced by the organism
Regulation	Responding to changes in the organism's surroundings
Reproduction	Making more organisms of the same kind

The Cell

Living things are made up of basic units called **cells**, each of which carries out life processes. Cells generally share some common structures. The **nucleus** controls cell activities. Surrounding the nucleus is a thick fluid called the **cytoplasm**, which is where most life processes occur. The cytoplasm is contained within the **cell membrane**—the "skin" of the cell—which controls the flow of materials into and out of the cell. Figure 1-1 shows a typical animal cell.

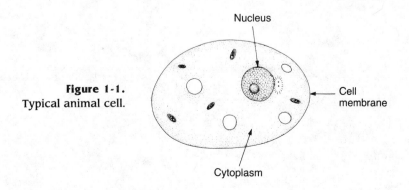

Figure 1-1.
Typical animal cell.

Nucleus

Cell membrane

Cytoplasm

Organisms and Their Environment

Living things interact constantly with their surroundings, called the *environment*. An organism's environment includes all living and nonliving things around it. Organisms obtain food, water, and oxygen from the environment. In turn, they release wastes, such as carbon dioxide. Thus, there is a continual exchange of materials between an organism and its environment.

Nutrition

Every organism needs food to stay alive. Food provides an organism with *nutrients*, which are used for growth and repair, and for producing energy. Some important nutrients are listed below in Table 1-2.

Table 1-2. Nutrients and Their Uses

Nutrient	*Use*
Proteins	Supply materials for growth and repair
Carbohydrates (sugars and starches)	Provide quick energy
Fats and oils	Provide stored energy
Vitamins	Assist life processes; prevent disease
Minerals	Supply materials for growth and repair; help carry out life processes

Green plants make their own food. Through a process called *photosynthesis* (Figure 1-2), plants use energy from sunlight to change carbon dioxide and water drawn from their environment into sugar. The sun's energy is thereby stored in the

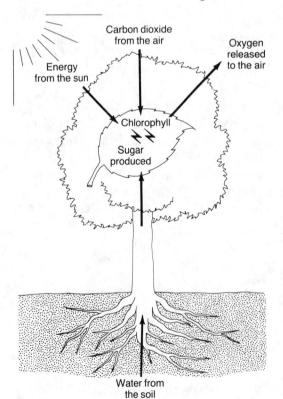

Figure 1-2. Photosynthesis.

sugar. Photosynthesis also produces oxygen, which is released into the environment. The green pigment *chlorophyll*, present in plant leaves, is necessary for this process to take place.

Animals obtain nutrients by eating plants or by eating other animals that feed on plants. The sugar in plants is used by animals to produce energy. The original source of energy in all food is the sun. Figure 1-3 shows one way to get food energy from the sun.

Figure 1-3. The energy in meat comes originally from the sun.

Respiration

Organisms use the energy stored in food through a process called **respiration**, which occurs in all cells. During respiration, nutrients in food combine with oxygen ("burn"). This chemical process releases energy and forms carbon dioxide and water as waste products.

Respiration is the opposite of photosynthesis, as shown below:

Photosynthesis: energy + carbon dioxide + water → sugar + oxygen

Respiration: sugar + oxygen → energy + carbon dioxide + water

All living things, including green plants, release energy from food through some form of respiration.

Water

Water is constantly exchanged between an organism and its environment. For example, we get water from the environment by eating and drinking, and we give water back by exhaling and perspiring.

Water is necessary for moving materials throughout an organism. For instance, blood, which is mostly water, carries nutrients to all parts of your body. Most of the chemical processes in living things can take place only in a watery environment. In addition, water is necessary for green plants to make food. For all of these reasons, life is not possible without water.

Reproduction

All living things come from other living things. **Reproduction** is the process by which an organism produces *offspring*—new individuals of its own kind. Each particular kind of organism is called a *species*. Lions are a species. Tigers are a different species. Since every individual organism eventually dies, reproduction ensures the continuation of its species.

There are two types of reproduction: *asexual* and *sexual*. **Asexual reproduction** involves only one parent. The offspring created are identical with the parent. Figure 1-4 shows examples of asexual reproduction.

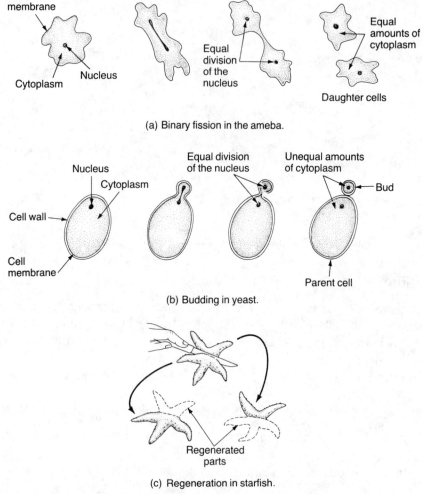

(a) Binary fission in the ameba.

(b) Budding in yeast.

(c) Regeneration in starfish.

Figure 1-4. Asexual reproduction: (a) binary fission;
(b) budding; (c) regeneration.

Sexual reproduction involves two parents, and produces offspring that are not identical with either parent. The female parent produces an egg cell, and the male parent produces a sperm cell. The joining together of these cells is called *fertilization*. The fertilized egg grows into a new individual.

Life Cycles

A puppy resembles an adult dog. A young elephant looks like a small version of its parents. However, this is not true of a frog or butterfly. Some organisms, like frogs and most insects, change so dramatically during their lives that the young may not resemble the adults at all. The changes an organism undergoes as it develops and then produces offspring make up its *life cycle*. Figure 1-5 illustrates the life cycles of a frog and a butterfly.

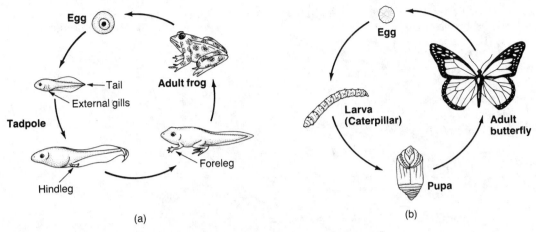

Figure 1-5. Life cycles: (a) frog; (b) butterfly.

EXERCISE 1

1. Which of the following does *not* describe a life process?
   (1) obtaining energy from food
   (2) responding to changes in the environment
   (3) production of new individuals
   (4) changing from a solid to a liquid

2. Which organism makes its own food?
   (1) a frog (2) a bird (3) a tree (4) a snake

3. The energy for photosynthesis comes from
   (1) the sun (2) oxygen (3) water (4) wind

4. The products of photosynthesis are sugar and
   (1) carbon dioxide (2) water (3) salt (4) oxygen

5. The process of "burning" food inside an organism's cells to release energy is called
   (1) excretion (2) photosynthesis (3) digestion (4) respiration

6. All of the following are life processes of animals *except*
   (1) digestion (2) respiration (3) photosynthesis (4) reproduction

7. Living things require sugar mainly for
   (1) growth (2) repair (3) energy (4) oxygen

8. The diagram shows an example of
   (1) spontaneous generation
   (2) asexual reproduction
   (3) photosynthesis
   (4) respiration

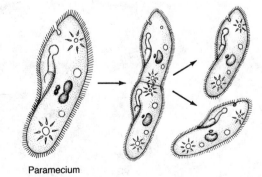

Paramecium

9. Which statement is true of all living things?
   (1) They have two parents.
   (2) They are exact copies of their parents.
   (3) The young look like small adults.
   (4) They come from other living things.

10. What is shown in the diagram below?
    (1) a life cycle
    (2) a food cycle
    (3) an ecosystem
    (4) a community

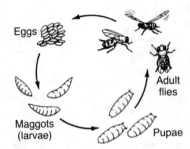

PROCESS SKILL: INTERPRETING AN EXPERIMENT

About three hundred years ago, the Italian scientist Francesco Redi wondered where maggots—small, wormlike organisms—come from. The popular belief at the time was that rotting meat turns into maggots. This idea, that living things could come from nonliving material, was called *spontaneous generation*. Redi designed an experiment to test this belief. He placed meat into eight jars. Four jars were left open; four were tightly sealed. Diagram 1 shows what Redi observed.

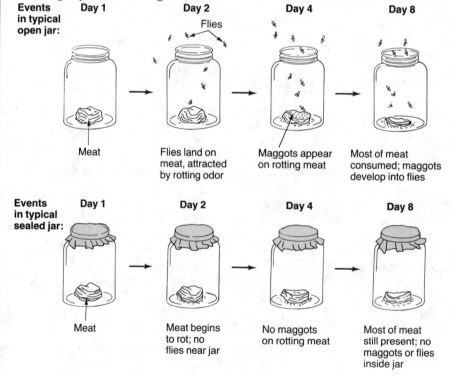

Diagram 1. Redi's first experiment: drawings show events in (*top*) a typical open jar; (*bottom*) a typical sealed jar.

PART II. LIVING THINGS AND THEIR ENVIRONMENT

Stimulus and Response

Living things must react to changes in their environment. For example, when the air gets too hot, you perspire. When the light gets too bright, the pupils of your eyes get smaller. Changes in the environment are called ***stimuli***. The ways in which living things react to these changes are called ***responses***. Table 1-3 gives examples of stimuli and responses for plants and animals.

Table 1-3. Stimuli and Responses

Organism	Stimulus	Response
Green plant	Light	Bends toward source of light
Dog	Food odor	Begins to salivate
Maple tree	Cold weather	Leaves change color
Human	Dust in nose	Sneezes

As you can see, no maggots appeared on the rotting meat in the sealed jars. However, not everyone was convinced that Redi's experiment had disproved spontaneous generation. Some people claimed that fresh air was needed for spontaneous generation to occur. Therefore, Redi performed a second experiment. This time the jars were covered by fine netting, which allowed fresh air into the jars but prevented flies from entering and landing on the meat. Diagram 2 shows what Redi observed in his second experiment. Study both diagrams and then answer the following questions.

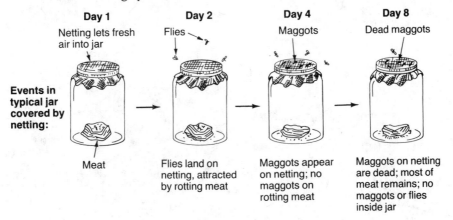

Diagram 2. Redi's second experiment.

1. Redi's second experiment provided clear evidence that the source of maggots was
   (1) air   (2) flies   (3) rotting meat   (4) either flies or air

2. Redi's experiments
   (1) supported the concept of spontaneous generation
   (2) did not support the concept of spontaneous generation
   (3) provided no evidence for or against spontaneous generation
   (4) proved that living things come from both living and nonliving objects

Some environments undergo extreme changes in temperature or other conditions during the year. Organisms that live in such environments have special responses that help them adjust to these changes.

1. Migration. Many birds that live in cold climates fly to warmer regions as winter approaches. This is called **migration**, moving from one environment to another.

2. Hibernation. Some animals, such as bears, survive the cold by sleeping for most of the winter. This is called **hibernation**.

3. Dormancy. Other living things may adjust to extreme environmental changes by entering a state of **dormancy**, becoming completely inactive. During the winter when a tree has lost its leaves, the tree is not dead—it is *dormant*. When spring comes, bringing warmer conditions, the tree grows new leaves. A seed may remain dormant for years, waiting for the proper conditions for growth.

An organism will survive only if it can adjust to changes in its environment.

Adaptations

Living things have special characteristics called *adaptations* that enable them to survive under a given set of conditions. Organisms may be adapted for life in water, soil, or air. For example, a fish has gills so it can breathe underwater. An earthworm's body shape helps it move through the soil. A bird has wings and light, hollow bones so it can fly. Many adaptations help an organism obtain food or escape predators in its environment. Figure 1-6 shows how certain birds are adapted for survival.

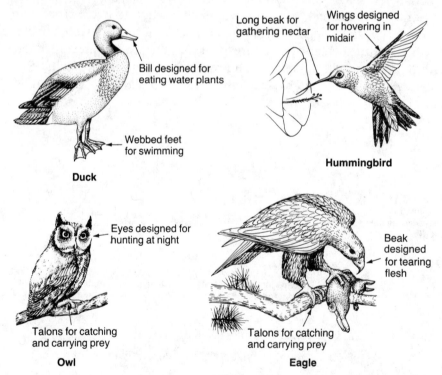

Figure 1-6. Adaptations of birds to their environment.

Earth's many environments include oceans, deserts, tropical rain forests, and the frozen Arctic tundra. Adaptations permit an organism to live in its own particular environment, or *habitat*. Organisms living in a dry, desert environment have adaptations that enable them to obtain and conserve water. For example, the cactus plant has an extensive root system that helps it reach what little water there is in the desert.

Animals living in the icy Arctic have adaptations that help them to endure the region's very cold temperatures. For instance, polar bears have thick coats of fur; seals and whales have layers of protective fat called *blubber*. Table 1-4 lists organisms from various habitats and their adaptations.

Table 1-4. Organisms and Their Adaptations

Organism	Habitat	Adaptation	Function
Giraffe	Grasslands	Long neck	Helps to reach leaves on trees, its main food
Arctic hare	Arctic	White fur in winter	Provides camouflage from predators
Monkey	Rain forest	Grasping tail	Acts as an extra hand, freeing hands and feet for other uses
Cactus	Desert	Waxy skin	Reduces water loss from evaporation

Communities and Ecosystems

A habitat usually contains many different types of organisms that interact with one another, and may depend on each other for survival. All the different organisms within a habitat make up a *community*. When you set up an aquarium containing plants, catfish, and guppies, you create a small community.

To set up an aquarium, you must provide more than just the fish and the plants. You need water, a source of oxygen, and light. You must also maintain a proper temperature. These nonliving factors together with the living members of the community make up an *ecosystem*.

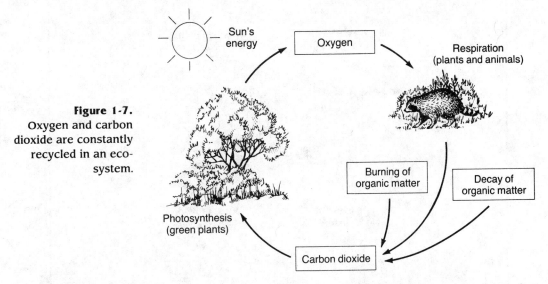

Figure 1-7. Oxygen and carbon dioxide are constantly recycled in an ecosystem.

The members of the community get the materials they need to survive from the ecosystem. In return, they give materials back, such as wastes and dead, decaying bodies.

Materials are constantly being recycled within an ecosystem. Figure 1-7 shows how oxygen and carbon dioxide are recycled. In our environment, oxygen is provided mainly by green plants, through photosynthesis. Energy, however, is not recycled and must be provided by an outside source, such as the sun.

PROCESS SKILL: INTERPRETING A TABLE

A student performed an experiment in which five plants were placed in sand and five plants were placed in soil. All ten plants were the same type, given equal amounts of water, and exposed to equal amounts of sunlight. The experiment lasted for two weeks. The table below shows the growth of each of the plants.

Plants in Soil	Increase in Height (in centimeters)	Plants in Sand	Increase in Height (in centimeters)
1	2.0	1	0.5
2	1.9	2	0.6
3	2.2	3	0.4
4	2.1	4	0.7
5	1.9	5	0.6

1. What conclusion may be drawn from this experiment?
   (1) Plants grow just as well in soil as in sand.
   (2) Plants grow taller in sand than in soil.
   (3) Plants grow taller in soil than in sand.

2. What might explain the results of this experiment?
   (1) Plants make their own food.
   (2) Plants produce oxygen.
   (3) Plants get more nutrients from soil than from sand.
   (4) Plants get more nutrients from sand than from soil.

3. Which bar graph correctly represents the *averaged* results of this experiment?

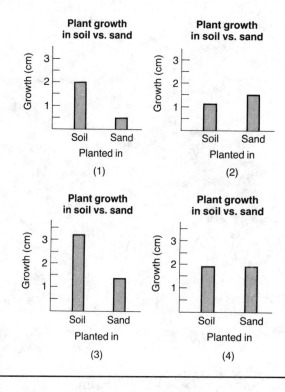

EXERCISE 2

1. A change in the environment is called a
   (1) stimulus (2) response (3) migration (4) reaction

2. A student touches a hot object and quickly pulls his hand away. This is an example of
   (1) a response followed by a stimulus (3) a response followed by a response
   (2) a stimulus followed by a response (4) a stimulus followed by a stimulus

3. Migration, hibernation, and dormancy are all methods of
   (1) producing food
   (2) changing the environment
   (3) adjusting to changes in the environment
   (4) producing energy

4. In which environment would you most likely find an animal with thick fur?
   (1) desert (2) tropical rain forest (3) Arctic (4) grassland

5. Life in the desert is difficult because there is very little
   (1) sunshine (2) water (3) oxygen (4) sand

6. Birds that are adapted to live in a watery environment would most likely have the type of feet shown in

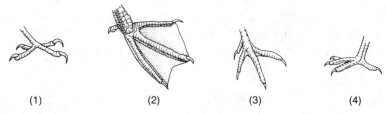

   (1)   (2)   (3)   (4)

7. All of the organisms that live in a pond make up
   (1) a habitat (2) a community (3) an environment (4) an ecosystem

Questions 8 and 9 refer to the diagram below, which represents a small ecosystem.

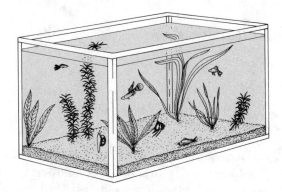

8. What is the main source of oxygen in this ecosystem?
   (1) the water (2) the fish (3) the snail (4) the green plants

9. The survival of this community depends upon a constant external supply of
   (1) energy (2) oxygen (3) carbon dioxide (4) plants

10. The original source of energy in our environment is
    (1) food (2) water (3) the sun (4) air

PART III. THE BALANCE OF NATURE

Food Chains

All organisms need energy to survive. They get this energy from nutrients in food. During photosynthesis, green plants produce sugar, our main source of food energy. Plant-eating animals obtain this energy-rich sugar when they eat and digest the plants.

Meat-eating animals also get energy from plants, but indirectly. For instance, when a lion eats a zebra, it obtains nutrients from the meat of the zebra. The lion gets its energy from these nutrients. The zebra had obtained these nutrients from the plants that are its food.

Both the lion and the zebra depend upon other organisms for their food. Green plants produce their own food. Therefore, green plants are called *producers*, while animals are called *consumers*. Every animal depends directly or indirectly on green plants for food and oxygen.

The nutrients in green plants get passed along from one organism to another in a sequence called a *food chain*. Grass produces food during photosynthesis. A zebra eats the grass to get its nutrients. A lion, in turn, eats the zebra.

When the lion dies, its body decays. Special organisms called *decomposers* break down the lion's remains and return nutrients to the soil. These nutrients can then be used again by plants, such as grass.

Decomposers include fungi, such as mushrooms and molds, and some bacteria. Fungi and bacteria cannot make their own food. They therefore depend on other living things for food. Decomposers are the last link in any food chain. Figure 1-8 shows an example of a food chain.

| Producers | Consumers | Decomposers |

Grasses   Rabbit   Hawk   Bacteria and fungi
in the soil

Figure 1-8. A food chain.

Suppose that, suddenly, there were no more zebras. How would the lion get its nutrients? A lion can also eat other animals. If something wiped out the zebra population, the lion would eat more of the other animals. Thus, the removal of one species, the zebra, would affect many other species.

Food Webs

Most ecosystems contain a number of food chains that are interconnected to form a *food web*, as shown in Figure 1-9. There is a delicate balance in an ecosystem among its producers, consumers, decomposers, and their environment. If this balance is disturbed, it may change the entire ecosystem.

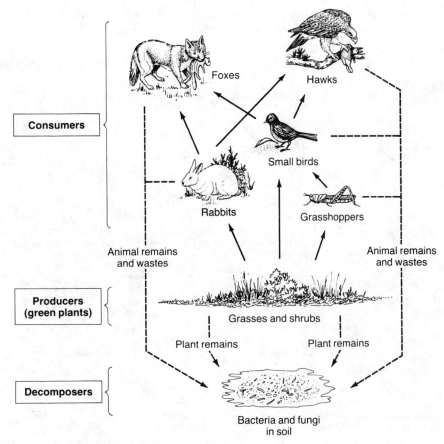

Figure 1-9. A food web consists of several interconnected food chains.

Upsetting the Ecosystem

Humans sometimes interfere with the balance of nature. For example, the early settlers in the northeastern United States killed off all the wolves in the region because the wolves preyed on farm animals. However, the wolves were the only natural enemies of the deer living in the area. Without the wolves to hold their numbers in check, the deer population increased to the point where many deer starved to death in winter.

The actions of people are not the only things that can disturb the balance of nature. Sometimes, the delicate balance may be upset by natural events such as floods and forest fires. In 1980, for instance, a volcano called Mount St. Helens, in the state of Washington, erupted violently. The explosion destroyed almost 100,000 acres of forest. In time, nevertheless, the forest will return to Mount St. Helens through a series of natural changes in the ecosystem.

Ecological Succession

After a forest fire or volcanic eruption has destroyed an ecosystem, the soil becomes enriched with minerals from the decaying remains of the plants and animals that once lived there. Soon, small new plants sprout. These become homes and food for insects and small animals. Eventually, these plants die and are replaced by other, larger plants. Each new community changes the environment, making it more suitable for the next community. Finally, a community emerges that is not replaced, called the *climax community*.

On Mount St. Helens, the climax community is the forest of spruce and fir trees that existed before the eruption, and the animals that inhabited it. The natural process by which one community is replaced by another in an orderly, predictable sequence is called *ecological succession*. Figure 1-10 illustrates the ecological succession of a barren area into a forest.

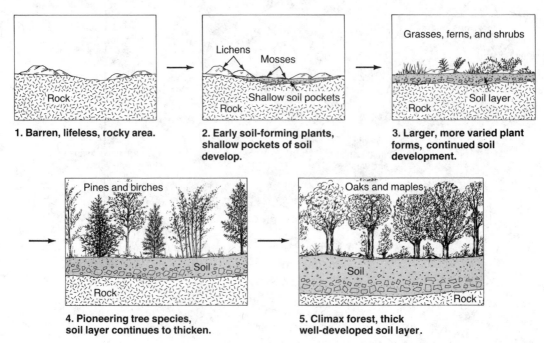

1. Barren, lifeless, rocky area.

2. Early soil-forming plants, shallow pockets of soil develop.

3. Larger, more varied plant forms, continued soil development.

4. Pioneering tree species, soil layer continues to thicken.

5. Climax forest, thick well-developed soil layer.

Figure 1-10. Ecological succession.

Conserving Natural Resources

A forest is an important natural resource. It supplies wood and oxygen, conserves soil and water, and provides a habitat for wildlife and recreation for people.

Forests that are destroyed can be replaced, although renewal takes a long time. This means the forest is a *renewable resource*, a resource that can be replenished. When plants and animals die and decay, they return nutrients to the soil. Therefore, soil is a renewable resource. Water, too, is a renewable resource, since it is constantly recycled through the environment (Figure 1-11).

Aluminum, like other minerals, is not replenished by nature. Minerals and other materials that are not naturally replaced are *nonrenewable resources*. To guarantee

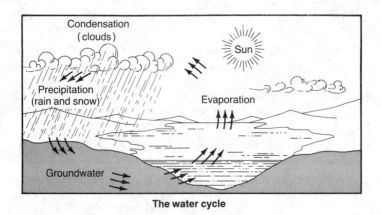

The water cycle

Figure 1-11. Water is constantly recycled through the environment.

an adequate supply of these valuable materials in the future, we must conserve and recycle them today.

Although nature does recycle water, soil, and forests, humans often use them up faster than nature can replace them. It is important, therefore, to conserve these resources as well, or we may have shortages of them someday.

EXERCISE 3

To answer questions 1 to 3, use the following diagram, which represents a food chain.

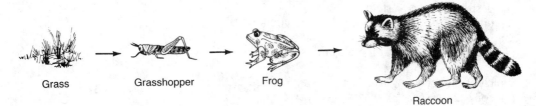

Grass   Grasshopper   Frog   Raccoon

1. Which organism is the producer in this food chain?
   (1) grass   (2) grasshopper   (3) frog   (4) racoon

2. In this food chain, the frog is a
   (1) producer   (2) consumer   (3) decomposer

3. Which type of organism is not shown in this diagram?
   (1) producer   (2) consumer   (3) decomposer

To answer questions 4 to 6 on page 16, use the diagram below of a food web.

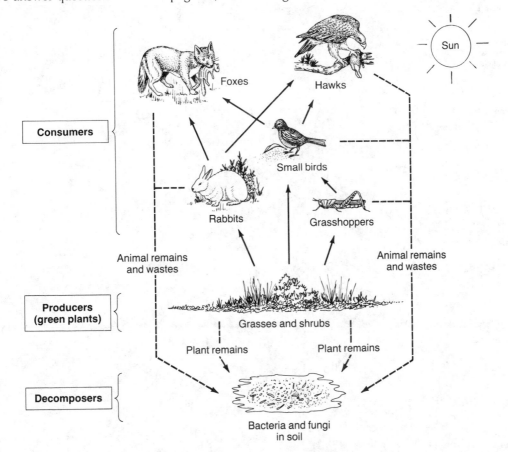

Sun

Consumers

Foxes   Hawks

Small birds

Rabbits   Grasshoppers

Animal remains and wastes   Animal remains and wastes

Producers (green plants)

Grasses and shrubs

Plant remains   Plant remains

Decomposers

Bacteria and fungi in soil

4. A decrease in the number of small birds would most likely result in an increase in the number of
   (1) rabbits   (2) hawks   (3) foxes   (4) grasshoppers

5. The organisms that return nutrients to the soil are the
   (1) producers   (2) consumers   (3) decomposers   (4) green plants

6. The producers in this ecosystem get their energy from
   (1) the rabbit   (2) the consumers   (3) the decomposers   (4) the sun

7. As the population of old shrubs decreases in a changing ecosystem, the population of new trees increases. The old community
   (1) destroys the ecosystem
   (2) prepares the ecosystem for the new community
   (3) is the climax community
   (4) does not provide nutrients to the soil

8. Which of the following is a nonrenewable resource?
   (1) soil   (2) silver   (3) water   (4) forest

Chapter 2. Living Systems: Human Systems

PART I. ORGANIZATION, SUPPORT, AND MOVEMENT OF THE BODY

Human Body Systems Are Interdependent

A human being is a complex organism, made up of a number of different *body systems*. Each system carries out a specific life process, and thereby contributes to the operation of the body as a whole.

In addition, all body systems are interdependent, and work with one another to keep a person alive. For instance, the respiratory system brings needed oxygen into the body; the oxygen is then transported throughout the body by the circulatory system.

Levels of Organization in the Human Body

1. Cells. Living things are made up of basic units called **cells**. The human body contains many types of cells, each designed to perform a different function. Figure 2-1 shows several kinds of cells.

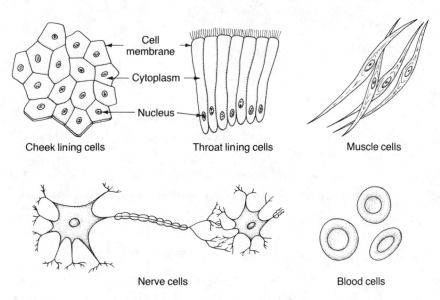

Cheek lining cells   Throat lining cells   Muscle cells

Cell membrane

Cytoplasm

Nucleus

Nerve cells   Blood cells

Figure 2-1. Different types of cells.

2. *Tissues.* A group of similar cells acting together forms a *tissue*. Skin tissue covers the body. Muscle tissue produces body movements. Table 2-1 lists some types of human tissues.

Table 2-1. Types of Human Tissue and Their Functions

Tissue	Function
Blood	Transports materials throughout the body
Bone	Supports and protects body and organs
Muscle	Helps body to move; aids in circulation, digestion, and respiration
Nerve	Carries messages
Skin	Covers and protects body; excretes wastes

3. *Organs.* A group of tissues working together forms an **organ**. The heart is an organ that pumps blood throughout the body. It is composed mainly of muscle tissue, but also contains blood tissue and nerve tissue. Table 2-2 lists some important organs.

Table 2-2. Important Organs and Their Functions

Organ	Function
Heart	Pumps blood
Kidney	Removes wastes from blood
Lung	Exchanges gases with the environment
Stomach	Breaks down food by physical and chemical means
Brain	Controls thinking and voluntary actions

4. *Organ Systems.* A group of organs acting together to carry out a specific life process makes up an **organ system**. For example, the circulatory system carries out the process of transport, moving materials throughout the body. Table 2-3 lists the human organ systems.

Table 2-3. Human Organ Systems

System	Function	Examples of Organs or Parts
Skeletal	Supports body, protects internal organs	Skull, ribs
Muscular	Moves organs and body parts	Arm and leg muscles
Nervous	Controls body activities; carries and interprets messages	Brain, spinal cord
Endocrine	Regulates body activities with hormones	Adrenal glands
Digestive	Breaks down food into a usable form	Stomach, intestines
Circulatory	Carries needed materials to body cells and waste materials away from cells	Heart, arteries, veins
Respiratory	Exchanges gases with the environment	Lungs, bronchi
Excretory	Removes wastes from the body	Kidneys, skin
Reproductive	Produces offspring	Ovaries, testes

The Skeletal System

The human *skeletal system*, shown in Figure 2-2, supports and protects the body and its organs. The skeletal system includes the *skull*, *spinal column*, *breastbone*, *ribs*, the bones of the *limbs* (arms and legs), and *cartilage*.

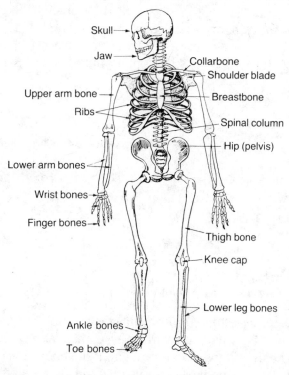

Figure 2-2. The human skeleton.

1. Bones and Cartilage. **Bones** are made of hard, strong material. **Cartilage** is a softer, more flexible tissue. Cartilage acts as a cushion between bones, and provides flexibility at the ends of bones. Disks of cartilage separate the bones of the spinal column, cushioning them from one another.

2. Joints. Where one bone meets another bone, a **joint** is formed. Most joints, such as the knee and elbow, allow the bones to move. However, some joints, like those in the skull, do not allow movement. Figure 2-3 shows three types of joints.

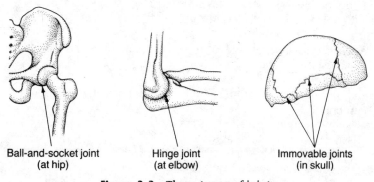

Ball-and-socket joint   Hinge joint   Immovable joints
(at hip)   (at elbow)   (in skull)

Figure 2-3. Three types of joints.

3. Ligaments and Tendons. At movable joints, the bones are held together by strips of tissue called *ligaments*. Bones are moved by muscles, which are attached to bones by *tendons*, cordlike pieces of tissue. A common sports injury is a torn Achilles tendon in the back of the lower leg.

The Muscular System

Muscles are masses of tissue that contract to move bones or organs. The *muscular system* contains two main kinds of muscles: *voluntary* and *involuntary*.

1. Voluntary Muscles. The skeletal muscles, which move bones, are examples of **voluntary muscles**—muscles that are controlled by our will. Skeletal muscles work with the skeleton to move body parts (see Figure 2-4), and thereby produce locomotion. *Locomotion* is the movement of the body from place to place. The muscles in the face and around the eyes are also voluntary muscles.

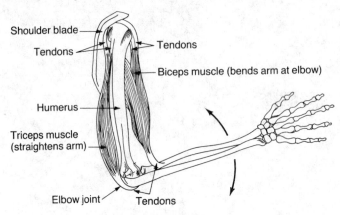

Figure 2-4. Muscles, tendons, and bones of the upper arm.

2. Involuntary Muscles. **Involuntary muscles** are not under our conscious control. There are two types of involuntary muscles: cardiac and smooth. *Cardiac* muscle, present only in the heart, pumps blood throughout the body. *Smooth* muscle, found in the respiratory, circulatory, and digestive systems, aids in breathing, controlling blood flow, and movement of food.

EXERCISE 1

---

1. A group of organs working together to carry out a life process is called
   (1) a cell   (2) a tissue   (3) an organ   (4) an organ system

2. A group of cells acting together makes up
   (1) a cell   (2) a tissue   (3) an organ   (4) an organ system

3. A tissue designed to carry messages throughout the body is most likely to be
   (1) skin   (2) muscle   (3) nerve   (4) bone

4. Going from the simple to the complex, which order correctly represents the organization of the human body?
   (1) organ system → organ → cell → tissue
   (2) cell → tissue → organ → organ system
   (3) tissue → cell → organ → organ system
   (4) organ → organ system → cell → tissue

5. Which body system supports and protects other body systems?
   (1) skeletal   (2) endocrine   (3) reproductive   (4) digestive

6. Which body system provides for movement of the body?
   (1) digestive   (2) circulatory   (3) muscular   (4) endocrine

7. Which group lists three parts of the skeletal system?
   (1) heart, stomach, brain   (3) bones, nerves, blood
   (2) tendons, nerves, brain   (4) cartilage, ligaments, bones

8. Which type of muscle is found only in the heart?
   (1) voluntary   (2) smooth   (3) cardiac   (4) involuntary

9. Which activity is most likely to be controlled by a smooth muscle?
   (1) breathing   (2) walking   (3) chewing   (4) thinking

10. The diagram below best demonstrates that

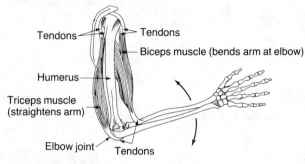

Tendons — Tendons
— Biceps muscle (bends arm at elbow)
Humerus
Triceps muscle (straightens arm)
Elbow joint — Tendons

   (1) the skeleton protects body organs
   (2) bones are held together at joints by ligaments
   (3) muscles and bones work together to move body parts
   (4) cartilage protects and cushions bones

PART II. REGULATION, DIGESTION, AND CIRCULATION

Regulation

The *nervous system* and the *endocrine system* work together to regulate body processes and actions. They provide us with a way of detecting and responding to stimuli.

The nervous system (Figure 2-5, page 22) is made up of the *brain*, *spinal cord*, *nerves*, and parts of the *sense organs*.

1. The **brain** receives and interprets *nerve impulses* ("messages"), and controls thinking, voluntary action, and some involuntary actions, such as breathing and digestion.

2. The **spinal cord** channels nerve impulses to and from the brain, and controls many automatic responses, or *reflexes*, such as pulling your hand away from a flame.

3. **Nerves** provide a means of communication between the sense organs, the brain and spinal cord, and muscles and glands.

4. The **sense organs**, which include the skin, eyes, ears, nose, and tongue, receive information from the environment.

Nerve cells, also called **neurons**, receive and transmit nerve impulses (see Figure 2-6, page 22). There are two kinds of neurons. *Sensory neurons* carry information from the sense organs to the brain or spinal cord. *Motor neurons* carry messages from the brain or spinal cord to muscles and glands, which respond to the messages.

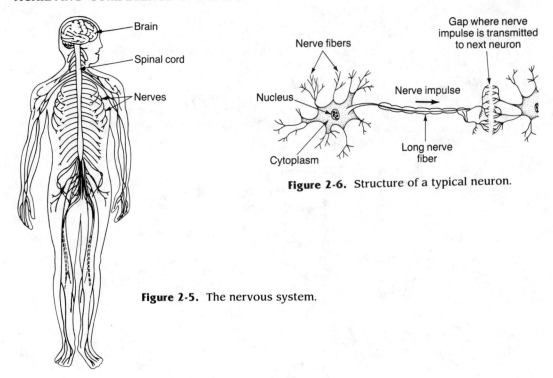

Figure 2-5. The nervous system.

Figure 2-6. Structure of a typical neuron.

The endocrine system is made up of glands. A **gland** is an organ that makes and *secretes* (releases) chemicals called **hormones**. Figure 2-7 shows some of the endocrine glands. When an endocrine gland secretes a hormone into the bloodstream, the blood carries the hormone to an organ, which responds in some way. For example, if you are suddenly faced with some danger, such as a snarling dog, the hormone *adrenaline* is released by your *adrenal gland*. The adrenaline makes your heart beat faster

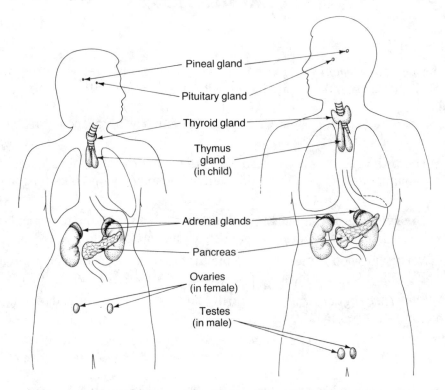

Figure 2-7. Some major glands of the endocrine system.

and your breathing more rapid. More sugar is released into your bloodstream to provide energy. These changes prepare your body to respond to the danger.

The Digestive System

Our cells need nutrients from food for energy, growth, and repair. The *digestive system* breaks down food into nutrients that can then be absorbed into the bloodstream and carried to the cells.

The digestive system, shown in Figure 2-8, consists of the digestive tract and the accessory organs.

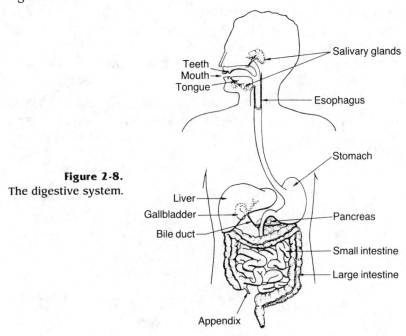

Figure 2-8.
The digestive system.

1. The *digestive tract* is a tube in which food travels through the body. It begins at the mouth and continues through the *esophagus, stomach, small intestine,* and *large intestine.*

2. The *accessory organs* are the *pancreas, gallbladder,* and *liver.* They produce digestive juices that are released into the digestive tract. Table 2-4 lists the digestive juices, where they are produced, and what foods they digest.

Table 2-4. Digestive Juices

Organ	Digestive Juice	Foods Acted On
Mouth	Saliva	Starches
Stomach	Gastric juice	Proteins
Small Intestine	Intestinal juices	Sugars, proteins
Pancreas	Pancreatic juice	Proteins, starches, fats
Liver	Bile	Fats

Note: Bile and pancreatic juice are secreted *by* the liver and pancreas *into* the small intestine, where digestion occurs.

The digestive system breaks down food by *physical* and *chemical* means. (1) Food is physically broken down into small bits by chewing, and by the action of muscles in the digestive tract. (2) The chemical breakdown of food into nutrients that can be used by cells is accomplished by chemicals called *enzymes*, found in the digestive juices.

Digestion starts in the mouth, and continues in the esophagus, stomach, and small intestine. When digestion has been completed, digested materials are absorbed into the bloodstream through the walls of the small intestine. Undigested materials, which make up the solid wastes called *feces*, pass on through the large intestine and are expelled from the body.

The Circulatory System

Nutrients absorbed into the blood must be transported to all body cells. This is the job of the *circulatory system*: to bring needed materials such as nutrients, water, and oxygen to the cells and to carry away wastes, like carbon dioxide, from the cells.

The components of the circulatory system are the *blood*, the *heart*, the *blood vessels* (*arteries, veins*, and *capillaries*), *lymph*, and *lymph vessels*.

1. Blood. The **blood** is a liquid tissue containing red and white blood cells, and platelets. The blood also carries dissolved nutrients, wastes, and hormones.

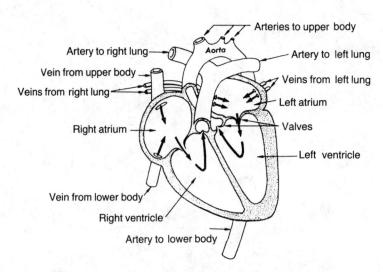

Figure 2-9. The heart.

2. Heart. The **heart** (Figure 2-9) is a muscle that contracts regularly to pump blood throughout the body. The blood is pumped from the heart to the lungs, where it receives oxygen and gets rid of carbon dioxide. The blood then returns to the heart to be pumped to the rest of the body, as shown in Figure 2-10.

3. Blood Vessels. The blood flows through a network of tubes called **blood vessels**. There are three types of blood vessels. **Arteries** carry blood away from the heart, while **veins** return blood to the heart. Connecting arteries to veins are the **capillaries**. Through these extremely thin blood vessels, essential materials are exchanged between the blood and the body's cells. Dissolved nutrients, water, and oxygen pass from the blood into the cells, and some wastes from the cells pass into the blood.

4. Lymph. Some of the watery part of the blood filters out through the walls of the capillaries into the surrounding tissue. This fluid, called **lymph**, bathes all the cells of the body. Lymph acts as a go-between in the exchange of materials between the blood and the cells. After receiving wastes from the cells, lymph is collected and returned to the bloodstream through **lymph vessels**.

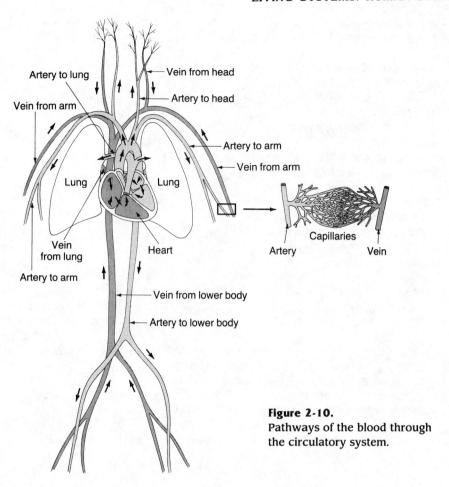

Figure 2-10.
Pathways of the blood through the circulatory system.

EXERCISE 2

---

1. Hormones are chemicals secreted by the
   (1) gall bladder   (2) brain   (3) endocrine glands   (4) small intestine

2. The endocrine system works with the nervous system to
   (1) digest nutrients
   (2) exchange gases with the environment
   (3) produce energy
   (4) regulate body activities

3. The brain, spinal cord, and sensory neurons are all part of the
   (1) nervous system   (3) circulatory system
   (2) respiratory system   (4) endocrine system

4. The human cell shown is designed to
   (1) store excess food
   (2) send and receive nerve impulses
   (3) cover and protect the body
   (4) carry oxygen to other cells

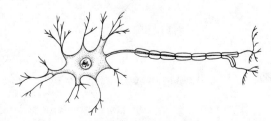

5. Food is broken down into a usable form by the
   (1) nervous system   (2) skeletal system   (3) digestive system   (4) circulatory system

6. Which group of structures lists parts of the digestive system?
   (1) heart, lungs, pituitary gland
   (2) adrenal gland, pituitary gland, thyroid gland
   (3) skin, kidneys, lungs
   (4) stomach, intestines, pancreas

7. Solid materials that are not digestible are eliminated from the body as
   (1) urine   (2) perspiration   (3) lymph   (4) feces

8. The function of the circulatory system is to
   (1) carry materials to and from the cells
   (2) break down food into a usable form
   (3) regulate body activities
   (4) respond to stimuli

9. Which group of structures all belong to the circulatory system?
   (1) heart, liver, and lungs   (3) arteries, kidneys, and stomach
   (2) arteries, veins, and capillaries   (4) bones, cartilage, and ligaments

10. Which represents the correct pathway of the nutrients in an apple once you take a bite?
    (1) circulatory system → cell → digestive system
    (2) cell → digestive system → circulatory system
    (3) digestive system → circulatory system → cell
    (4) circulatory system → digestive system → cell

Questions 11 and 12 refer to the four diagrams of organ systems shown below.

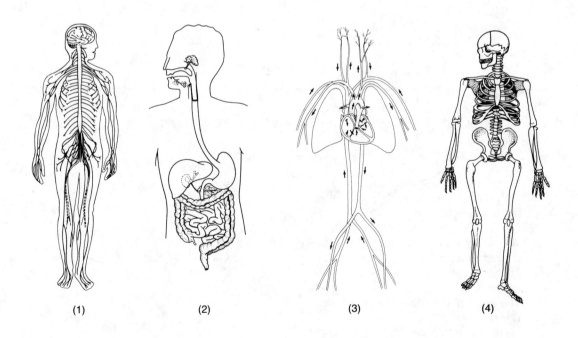

(1)   (2)   (3)   (4)

11. Which diagram represents the digestive system?

12. Which system transports needed nutrients to all the cells of the body?

PART III. RESPIRATION, EXCRETION, AND REPRODUCTION

The Respiratory System

The circulatory system provides oxygen to the cells. Cells use this oxygen in the process of **cellular respiration**, in which nutrients from digested food combine with the oxygen to release energy and produce the waste materials carbon dioxide and water. This chemical process takes place in all body cells.

The *respiratory system*, illustrated in Figure 2-11, brings oxygen from the air to the blood, and returns carbon dioxide from the blood to the air. This process is called **respiration**.

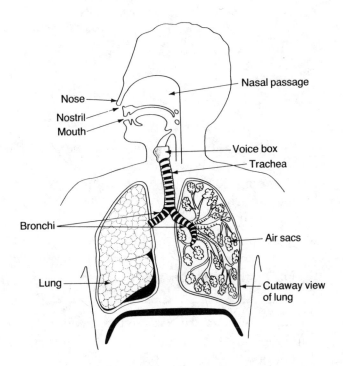

Figure 2-11. The respiratory system.

When you breathe in (*inhale*), air enters the nose or mouth and passes through the **trachea**, or windpipe. The trachea branches off to each lung through tubes called **bronchi**. The lungs contain millions of tiny *air sacs*, surrounded by capillaries. Here, respiratory gases are exchanged—oxygen enters the blood while carbon dioxide leaves the blood and is breathed out (*exhaled*).

The oxygen that enters the blood is carried to the cells of the body, where an exchange of gases again takes place. This time, oxygen leaves the blood and enters the cells, while carbon dioxide leaves the cells and goes into the blood. The carbon dioxide is returned to the lungs to be exhaled. This process is repeated constantly.

The Excretory System

The activities of the body's cells produce waste materials that must be removed. These wastes are eliminated from the blood, and, eventually, from the body, by the *excretory system*.

The excretory system consists of the *lungs, skin, kidneys,* and *liver*.

1. The *lungs* expel the waste products carbon dioxide and water vapor from your body each time you exhale.

2. The *skin* expels wastes when you perspire. Microscopic sweat glands deep in the skin excrete *perspiration*, a liquid waste consisting mostly of water and salts. Perspiration leaves the body through the *pores*, which are tiny openings in the surface of the skin (Figure 2-12).

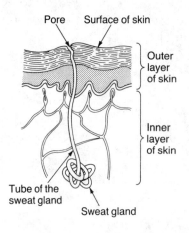

Figure 2-12.
Sweat glands in the skin expel wastes from the body through pores.

3. The two *kidneys* (Figure 2-13) help to maintain the proper balance of water and minerals in the body. As blood flows through the kidneys, excess water, salts, urea, and other wastes are removed from the blood. These substances make up a fluid called *urine*. Urine is sent through a tube from each kidney to the *bladder*, where it is stored until excreted from the body.

PROCESS SKILL: INTERPRETING A DIAGRAM

The diagram below is a *schematic* representation of the circulatory system. In other words, it is not meant to be a realistic drawing of body parts, but only to show the basic scheme of the system—the relationships among its parts and the sequence of events that occur in the system.

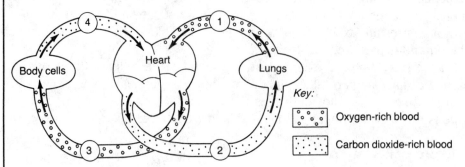

The circulation of blood is vital to the process of respiration, since the blood carries fresh oxygen to the cells of the body and returns carbon dioxide to the lungs to be expelled.

4. The *liver* produces *urea*, a waste resulting from the breakdown of proteins. Urea is taken by the blood to the kidneys and expelled from the body in urine. The liver also removes harmful substances from the blood.

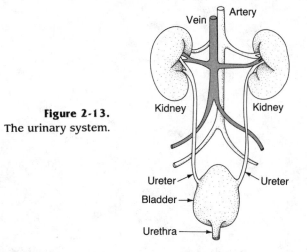

Figure 2-13.
The urinary system.

The Reproductive System

The reproductive system is responsible for the production of offspring. There are two human reproductive systems, male and female, as shown in Figure 2-14 on page 30.

1. Male. The male reproductive system consists of the *testes*, *penis*, and *sperm ducts*. The **testes** produce *sperm cells*, the male sex cells. During reproduction, these

As you have learned, arteries are blood vessels that carry blood away from the heart. Which blood vessels in the diagram are arteries? The arrows indicate that blood vessels 2 and 3 carry blood away from the heart, so they are arteries. Blood vessels 1 and 4, which return blood to the heart, are veins. Study the diagram and then answer the following questions.

1. Blood rich in oxygen is found in blood vessels
   (1) 1 and 2   (2) 2 and 3   (3) 1 and 3   (4) 2 and 4

2. Compared with blood vessel 1, the amount of carbon dioxide in blood vessel 2 is
   (1) greater   (2) less   (3) the same

3. Which statement is true?
   (1) All arteries carry oxygen-rich blood.
   (2) All veins carry oxygen-rich blood.
   (3) Arteries from the heart to the lungs carry oxygen-rich blood.
   (4) Veins from the lungs to the heart carry oxygen-rich blood.

cells pass through tubes called *sperm ducts*, where they mix with a fluid to form *semen*. The semen is delivered through the *penis* into the female's reproductive system.

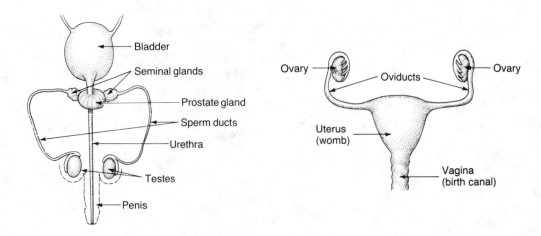

Figure 2-14. Human reproductive systems: (*left*) male; and (*right*) female.

2. Female. Making up the female reproductive system are the *ovaries, oviducts, uterus, vagina,* and *mammary glands*. The *ovaries* produce *egg cells,* the female reproductive cells. Once a month, an egg cell leaves an ovary and travels through one of the *oviducts* to the *uterus,* or womb. If sperm cells are present in the oviduct, fertilization may take place.

After fertilization has occurred, the fertilized egg attaches itself to the inner wall of the uterus. There it develops into a new offspring over a period of about nine months. At the end of this time birth takes place, and the offspring emerges through the *vagina,* or birth canal. The newborn baby may be fed milk produced by the mother's *mammary glands,* or breasts.

EXERCISE 3

1. The process by which energy is released from nutrients is called
   (1) cellular respiration   (2) excretion   (3) digestion   (4) circulation

2. Where does cellular respiration take place?
   (1) in the blood only   (2) in the lungs only   (3) in the heart only   (4) in all body cells

3. The respiratory system includes the
   (1) heart, liver, lungs
   (2) lungs, trachea, nose
   (3) stomach, esophagus, liver
   (4) heart, arteries, veins

4. Which represents the correct order in which oxygen enters the body?
   (1) nose, trachea, bronchi, lungs
   (2) bronchi, nose, trachea, lungs
   (3) lungs, bronchi, trachea, nose
   (4) nose, bronchi, trachea, lungs

5. The exchange of gases between the air and the blood takes place in the
   (1) nose   (2) trachea   (3) bronchi   (4) lungs

6. At each body cell,
    (1) carbon dioxide enters the blood, and oxygen leaves the blood
    (2) both carbon dioxide and oxygen enter the blood
    (3) both carbon dioxide and oxygen leave the blood
    (4) oxygen enters the blood and carbon dioxide leaves the blood

7. Which organ belongs to both the excretory system and the respiratory system?
    (1) heart   (2) kidney   (3) lungs   (4) liver

8. The excretory system includes the
    (1) kidneys, liver, lungs
    (2) lungs, trachea, nose
    (3) stomach, esophagus, liver
    (4) heart, arteries, veins

9. Which of the following helps remove wastes from the body?
    (1) the skull   (2) the skin   (3) the spinal cord   (4) the stomach

Use the diagrams below to answer questions 10 and 11.

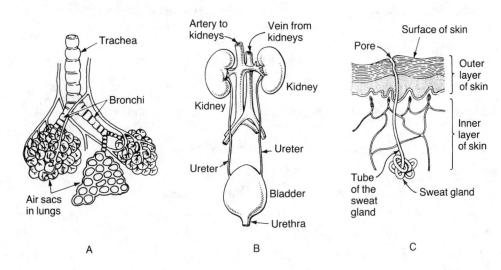

10. To which system do all three structures belong?
    (1) excretory   (2) circulatory   (3) respiratory   (4) skeletal

11. Which diagram may also represent part of the respiratory system?
    (1) *A*   (2) *B*   (3) *C*   (4) all of these

12. Male is to sperm, as female is to
    (1) egg   (2) testes   (3) oviduct   (4) uterus

Chapter 3. Living Systems: Microorganisms

PART I. CELLS

Cells and Life Processes

All living things are made up of one or more *cells*. For instance, an ameba consists of a single cell, while a human consists of billions. Each cell carries out basic life processes. Smaller structures within the cell perform these life processes. Table 3-1 lists some of these structures.

Table 3-1. Some Cell Structures and Their Functions

Structure	Function
Mitochondria	Respiration—where food is "burned" (combined with oxygen) to produce energy.
Ribosomes	Synthesis—where proteins are made.
Lysosomes	Digestion—where digestive enzymes are stored.
Nucleus	Reproduction—where genetic material is stored.
Vacuole	Digestion and excretion—where digestion occurs or where excess fluid is stored.
Chloroplasts (present in plant cells only)	Photosynthesis—where *glucose* (sugar) is produced in green plants.

Plant cells and animal cells have many structures in common, but they also have differences (see Figure 3-1).

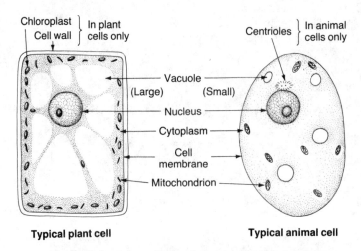

Figure 3-1. Comparison of plant and animal cells.

1. Plant cells. In plant cells, a *cell wall* encloses the entire cell, including the cell membrane. The tough cell wall gives support to the plant's structure. Found only in green plant cells are structures called *chloroplasts*, which contain chlorophyll. Chloroplasts are the site of photosynthesis, the food-making process of plants.

2. Animal cells. Animal cells do not have cell walls, but are enclosed only by the cell membrane. Found only in animal cells are the *centrioles*, which participate in cell division.

Cell Division

All cells come from other cells through the process of **cell division**. In this process, one "parent" cell divides into two new "daughter" cells. Every parent cell passes along to its daughter cells a set of "operating instructions" necessary for the cells to function properly. This *genetic information* is contained in threadlike structures called *chromosomes*, found in the cell nucleus.

The genetic information in the chromosomes also gives the cell, or the organism it belongs to, its individual characteristics or *traits*, like size and shape. All members of a given species have the same number of chromosomes in each body cell. Chromosomes and their genetic information are passed on to the next generation during reproduction.

One-celled organisms reproduce through a kind of cell division called *mitosis*. In this process, a cell divides into two identical daughter cells, each of which contains the same number of chromosomes as the original parent cell (see Figure 3-2). When a one-celled organism undergoes mitosis, each new cell produced is a complete new organism.

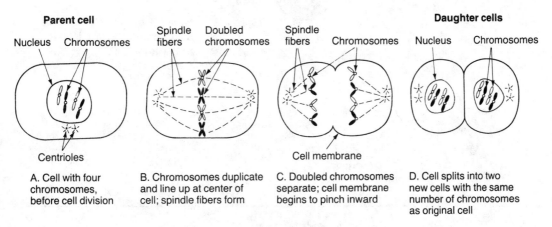

Figure 3-2. Cell Division: Mitosis produces two new cells with the same number of chromosomes as the original cell.

In an organism made up of many cells, cells must duplicate themselves to build new tissue for growth and to repair damaged tissue. They do this through mitosis. Some of these organisms also produce offspring by mitotic cell division. In fact, any organism that reproduces *asexually* (with just one parent) does so through mitosis. The offspring created are genetically identical to the parent.

Sexual Reproduction

Some organisms reproduce *sexually*, with two parents. Sexual reproduction involves the joining of two special reproductive cells, one from each parent. These *sex cells* (sperm cells and egg cells) have only half the number of chromosomes held by other body cells. Sex cells are formed by a type of cell division called *meiosis*.

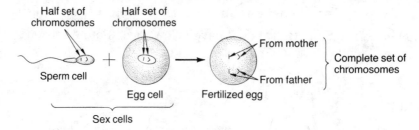

Figure 3-3. Fertilization.

During sexual reproduction, a sperm cell from the male parent joins with an egg cell from the female parent. This is called *fertilization*. Since each sex cell contains half of a normal set of chromosomes, when they join, they form one cell with a complete set of chromosomes (Figure 3-3). This cell then develops into a new organism through mitosis. The new organism is not identical to either parent, but has traits from both. In this way, sexual reproduction leads to variation in the next generation. Figure 3-4 shows a possible result of sexual reproduction in chickens.

Figure 3-4.
Sexual reproduction leads to variation in the offspring.

EXERCISE 1

1. The basic unit of all living things is the
   (1) nucleus   (2) cell   (3) organ   (4) chromosome

2. What do all living things have in common?
   (1) They reproduce sexually.
   (2) They manufacture their own food.
   (3) They are made up of one or more cells.
   (4) They are identical to their parents.

3. What do all cells have in common?
   (1) They carry out life processes.   (3) They take part in fertilization.
   (2) They reproduce sexually.   (4) They are unable to reproduce.

4. Cell division is necessary for all of the following *except*
   (1) growth   (2) repair of damaged tissue   (3) reproduction   (4) cellular respiration

5. Variation in a new generation of organisms is the result of
   (1) sexual reproduction involving one parent
   (2) sexual reproduction involving two parents
   (3) asexual reproduction involving one parent
   (4) asexual reproduction involving two parents

6. Which of the following diagrams shows sexual reproduction?

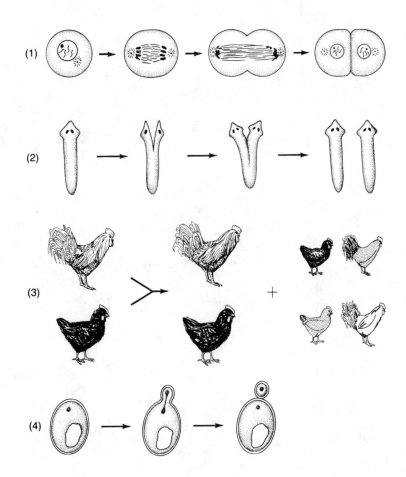

PART II. MICROORGANISMS

Harmful Microorganisms

A *microorganism* is a very small organism that usually cannot be seen without a *microscope*. This instrument makes small objects viewed through it appear larger. Several kinds of microorganisms are pictured in Figure 3-5, page 36, as they would appear when seen through a microscope.

Certain microorganisms, sometimes called *disease germs*, can be harmful to humans and other living things. For example, an **infectious disease** is an illness caused by microorganisms that can be transmitted, or passed on, from one individual to another. Table 3-2 lists some infectious diseases and the types of microorganisms that cause them.

Table 3-2. Some Infectious Diseases and Their Causes

Disease	Type of Microorganism
Pneumonia Strep throat Botulism	Bacteria
Common cold Flu (influenza) Mono (infectious mononucleosis) AIDS	Viruses
Athlete's foot	Fungus
Amebic dysentery	Protozoan

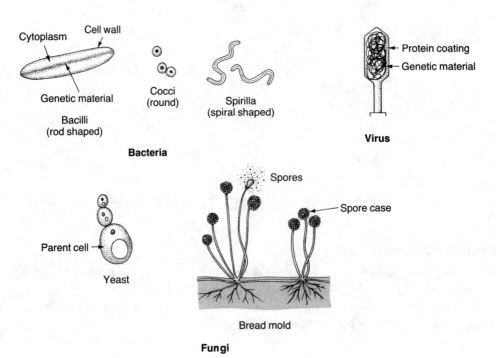

Figure 3-5. Different types of microorganisms.

Preventing Infection

When a harmful microorganism enters another organism and reproduces, an *infection* results. *Contamination* and *spoilage* of food are other serious problems caused by microorganisms. Approaches to controlling microorganisms include removing them, killing them, or preventing them from growing. There are a number of ways to defend ourselves against disease germs.

1. Sterilization kills all microorganisms in an area. The most common method used is to apply intense heat. For instance, a surgeon's instruments are *sterilized* (made free of disease-producing germs) by heating the instruments.

2. Disinfection is the destruction of all or most of the harmful microorganisms in an area by using chemicals called *germicides*.

3. Pasteurization is the process of heating milk or other foods to kill bacteria that cause spoilage and disease.

4. Canning is another method of preserving foods. In this process, foods are sealed inside airtight containers, which are then sterilized by heating. Canning prevents microorganisms from multiplying in the food.

5. Refrigeration slows the growth rate of bacteria that spoil food. For example, the bacteria that cause milk to turn sour grow 30 times faster at room temperature than they do in a refrigerator. However, refrigeration does not kill the bacteria.

6. Freezing food slows the growth of bacteria even further, but this method has the disadvantage of altering the taste and texture of some foods.

Noninfectious Diseases

Microorganisms are not the cause of every disease. Arthritis, high blood pressure, asthma, and cancer are examples of **noninfectious diseases**, which cannot be transmitted from one individual to another.

Causes of noninfectious diseases include poor diet, malfunctioning endocrine glands, damaged organs, allergies to foreign substances, and reactions to chemicals or radiation in the environment. Table 3-3 lists some noninfectious diseases and their causes.

Table 3-3. Some Noninfectious Diseases and Their Causes

Disease	Cause
Scurvy	Vitamin C deficiency
Anemia	Iron deficiency
Hay fever	Allergy to pollen
Gastric ulcer	Excess stomach acid
Skin cancer	Radiation from the sun
Lung cancer	Cigarette smoke or other cancer-causing agents
Diabetes	Malfunctioning pancreas

Helpful Microorganisms

Only about five percent of the known microorganisms are harmful. In fact, many microorganisms are beneficial to us and to other living things, and some are even essential to our well-being. Helpful microorganisms include the bacteria of decay, or *decomposers*, which break down dead organisms and return their nutrients to the environment.

Some microorganisms are used to help produce certain foods. Yogurt, for example, is produced by the action of bacteria on milk. *Molds*, a type of fungus, help in making cheese. *Yeasts* are fungi that cause bread to rise and produce alcohol in beer and wine. Humans even need certain kinds of bacteria inside their bodies to aid in digestion.

PROCESS SKILL: DESIGNING A CONTROLLED EXPERIMENT; EXPLAINING EXPERIMENTAL TECHNIQUE; MAKING PREDICTIONS

In the late nineteenth century, the French scientist Louis Pasteur investigated a disease called *anthrax*, which was killing sheep and cattle. He suspected that the disease was caused by a particular kind of bacteria. To test this *hypothesis*, or educated guess, he performed an experiment.

Pasteur heated a sample of the bacteria just enough to weaken, but not kill, them. Using a group of 50 sheep, he injected 25 sheep with the weakened bacteria, and left the other 25 sheep alone. The injected sheep became slightly ill, but soon recovered. Several weeks later, Pasteur injected all 50 sheep with a large dose of healthy bacteria, strong enough to kill a normal sheep. After a few days, all 25 sheep that had been injected with weakened bacteria were still alive, while the other 25 were all dead of anthrax.

Through this experiment, Pasteur demonstrated that anthrax was in fact caused by bacteria. He also showed that by giving sheep a mild case of the disease, he could protect them from more serious infection in the future. This procedure is called *immunization*, and is used today to protect humans from many infectious diseases.

Pasteur's demonstration was effective because he followed the scientific method in designing his experiment. Why was it necessary to use two groups of sheep? The group that was made immune by the first injection was the *experimental group*. To be sure it was the injection that made them immune, another group was needed for comparison. This was the *control group*. Both groups had to be treated exactly the same, except for the condition that was being tested. The condition that was different (the immunization) was the *variable*.

*   *   *

A teacher asked her students to perform an experiment on factors that affect the souring of milk. Betty obtained three containers of "Surefresh" milk. One she kept at room temperature, 20°C. Another she refrigerated at 5°C, and the third she kept at 1°C. She checked the milk every day. Examine her results in the table below, and answer the questions that follow on the facing page.

Temperature	Time Until Milk Turned Sour
20°C	1 day
5°C	3 days
1°C	7 days

1. The variable in this experiment was the
   (1) brand of milk   (2) amount of light
   (3) temperature at which the milk was kept

2. Betty might reasonably conclude that as the temperature increases, the length of time that milk takes to turn sour
   (1) increases   (2) decreases   (3) remains the same

3. Based on her experiment, Betty should predict that at 10°C, the milk would turn sour in about
   (1) 1 day   (2) 2 days   (3) 5 days   (4) 9 days

Marshall decided to compare the souring times of three different brands of milk. He left a container of "Surefresh" milk at 20°C, one of "Sunshine" milk at 5°C, and one of "Dairytime" milk at 1°C. All three containers had the same expiration date stamped on them, and all were fresh when the experiment began. Here are his results:

Brand	Temperature	Time to Sour
Surefresh	20°C	1 day
Sunshine	5°C	2 days
Dairytime	1°C	8 days

4. Which conclusion could Marshall reasonably draw from his experiment?
   (1) "Surefresh" milk turns sour faster than "Sunshine."
   (2) "Sunshine" milk turns sour faster than "Dairytime."
   (3) All brands of milk are the same.
   (4) He couldn't conclude any of these.

5. The major mistake that Marshall made in his experiment was that he
   (1) used too many different brands of milk
   (2) had too many variables
   (3) didn't make sure that the milk was fresh at the start of the experiment

Marshall's teacher told him to check all three brands at the same temperature. When he did this, at 5°C, he got the following results:

Brand	Time to Sour
Surefresh	3 days
Sunshine	2 days
Dairytime	2 days

All three containers had the same expiration date, and were fresh before the experiment began. Marshall concluded that he should always buy "Surefresh." Other students, however, repeated his experiment and found that, on the average, there was no difference among the three brands.

6. What was wrong with Marshall's conclusion?
   (1) He didn't measure the time accurately enough.
   (2) 20°C is too warm to store milk.
   (3) He should have tried several different temperatures.
   (4) He should have repeated the experiment several times to see if he always got the same results.

EXERCISE 2

1. Infectious diseases are caused by
   (1) microorganisms
   (2) allergies
   (3) poor diet
   (4) chemicals in the environment

2. Which disease is a type that can be transmitted from one individual to another?
   (1) hay fever   (2) the flu   (3) skin cancer   (4) diabetes

3. Noninfectious diseases can be caused by all of the following *except*
   (1) malfunctioning glands
   (2) damaged organs
   (3) bacteria
   (4) reaction to chemicals in the environment

4. Scurvy is a disease that sailors often used to get on long voyages. It was found that eating oranges and limes prevented scurvy. This suggests that scurvy is a disease caused by
   (1) a microorganism   (2) a deficiency in diet   (3) an allergy   (4) a damaged organ

5. Which of the following methods may be used to kill harmful bacteria in food?
   (1) freezing and refrigerating
   (2) wrapping and bottling
   (3) wrapping and freezing
   (4) pasteurizing and sterilizing

6. Disinfection is a method used to
   (1) control infectious diseases
   (2) increase the number of bacteria
   (3) increase the rate of spoilage
   (4) spread diseases

7. Studies have determined that smoking cigarettes can cause lung cancer. This is an example of a disease caused by
   (1) reaction to chemicals in the environment
   (2) microorganisms
   (3) deficiencies in diet
   (4) bacteria

8. Lyme disease is an infectious disease transmitted to humans by a deer tick (a small insect). The numbers in the diagram below show how many cases of Lyme disease were reported in 1987 and 1988 in several northeastern states. Which state had the second-most cases of Lyme disease?
   (1) New York
   (2) New Jersey
   (3) Rhode Island
   (4) Connecticut

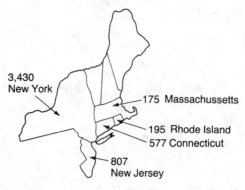

Northeastern U.S.

3,430 New York

175 Massachussetts

195 Rhode Island

577 Connecticut

807 New Jersey

Chapter 4. Earth's Changing Surface

PART I. EARTH'S SURFACE

Surface Changes

The surface of planet Earth is constantly undergoing change. For example, new mountains are rising as old mountains are being worn away. Rivers are carving valleys deep into the face of the land. Large rocks are being broken down into smaller and smaller rocks. There are two sets of forces, acting in opposite ways, that shape Earth's surface. External forces wear the surface down, while internal forces build the surface up.

Surface Materials

Earth's rocky outer layer is called the *crust*. The surface of the crust consists of bedrock, rock fragments, and soil, as shown in Figure 4-1.

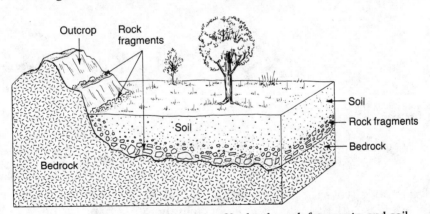

Figure 4-1. Earth's surface consists of bedrock, rock fragments, and soil.

Bedrock is the solid rock portion of the crust. Bedrock that becomes exposed at Earth's surface is called an *outcrop*. Rock fragments are pieces of broken-up bedrock. They can range in size from giant boulders to tiny grains of sand.

Soil is a mixture of small rock fragments and *organic matter* (materials produced by living things, such as decaying leaves and animal wastes). Soil and fragments of rock make up most of Earth's surface, with the bedrock hidden underneath.

Minerals

Rocks are composed of *minerals*, which are naturally occurring solid substances made of inorganic (nonliving) material. Of the many minerals found in the rocks of Earth's crust, feldspar is the most abundant. Some other common minerals are quartz, mica, and calcite. Minerals have certain *physical* and *chemical properties* by which they can be identified.

1. **Physical properties** of minerals include hardness, luster, cleavage, and color.

Hardness is the resistance of a mineral to being scratched. Minerals are assigned a number between 1 and 10 to indicate their hardness, with 1 being the softest and 10 the hardest. The hardness scale shown in Table 4-1 lists the minerals used as reference points. A mineral can be scratched only by another mineral with a higher number on the hardness scale.

Table 4-1. Hardness Scale of Minerals

Mineral	Hardness	Mineral	Hardness
Talc	1	Feldspar	6
Gypsum	2	Quartz	7
Calcite	3	Topaz	8
Fluorite	4	Corundum	9
Apatite	5	Diamond	10

Luster refers to how a mineral looks when it reflects light. A mineral can look metallic, glassy, greasy, or earthy.

Cleavage is a mineral's tendency to break along smooth, flat surfaces. The number and direction of these surfaces are clues to a mineral's identity. Cleavage often causes a mineral to break into characteristic shapes, as shown in Figure 4-2. Not all minerals have definite cleavage; some fracture unevenly when broken.

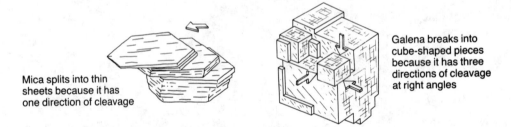

Galena breaks into cube-shaped pieces because it has three directions of cleavage at right angles

Mica splits into thin sheets because it has one direction of cleavage

Figure 4-2. Cleavage in (*left*) mica; (*right*) galena.

Color is not always a reliable guide to a mineral's identity. Various samples of the same mineral may have different colors. On the other hand, samples of different minerals may share the same color. Color is best used together with other properties to identify a mineral.

2. Minerals also have **chemical properties**, such as how they react with an acid. For example, calcite, the chief mineral in limestone and marble, fizzes when hydrochloric acid is placed on it. The fizzing is caused by a chemical reaction between the calcite and the acid in which bubbles of carbon dioxide gas are given off.

Rocks

The **rocks** that form Earth's crust are natural, stony materials composed of one or more minerals. Like minerals, rocks are identified by their physical and chemical properties. Rocks are classified into three groups—igneous, sedimentary, and metamorphic—depending on how they are formed.

1. **Igneous rocks** are produced by the cooling and hardening of hot, liquid rock. This melted rock material is called *magma* when underground and *lava* when it pours onto Earth's surface. Different igneous rocks are generally identified by their color and by the size of the mineral grains (crystals) they contain.

Igneous rocks that form from rapid cooling of lava, called *volcanic rocks*, contain tiny crystals. Basalt is a dark-colored volcanic rock composed of crystals too small to be seen with the unaided eye.

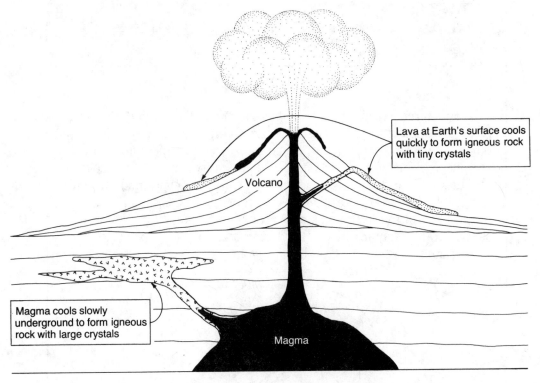

Lava at Earth's surface cools quickly to form igneous rock with tiny crystals

Volcano

Magma cools slowly underground to form igneous rock with large crystals

Magma

Figure 4-3. Formation of igneous rocks.

Igneous rocks that form underground by slow cooling of magma develop large crystals. Granite is a light-colored igneous rock that contains large, easily visible mineral grains. Figure 4-3 shows processes that produce igneous rocks.

2. **Sedimentary rocks** form from particles called *sediments* that pile up in layers. These sediments may be small rock fragments or seashells. Sedimentary rocks usually form underwater. For example, when a stream carrying particles of sediment empties into an ocean or lake, the particles settle to the bottom in layers. Eventually, these

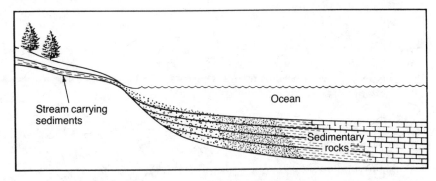

Stream carrying sediments

Ocean

Sedimentary rocks

Figure 4-4. Sedimentary rocks form in layers.

layers harden into sedimentary rock (see Figure 4-4 on page 43). Table 4-2 lists some common sedimentary rocks, what they are made of, and where they form.

3. *Metamorphic rocks* are produced when either igneous or sedimentary rocks undergo a change in form caused by heat, pressure, or both. This can take place when magma heats rocks it comes in contact with or when forces deep underground squeeze rocks for long periods of time. The high temperatures and pressures thus

---

PROCESS SKILL: DETERMINING QUALITATIVE RELATIONSHIPS

The Rock Cycle

The three types of rocks—*igneous*, *sedimentary*, and *metamorphic*—are subject to processes that can change any one type into another type. Igneous rocks can be changed into sedimentary or metamorphic rocks by various processes. Sedimentary rocks can be recycled into new sedimentary rocks, or changed into igneous or metamorphic rocks. Likewise, metamorphic rocks can become igneous or sedimentary rocks. All these changes and processes make up the *rock cycle*, shown in the diagram below. Study the diagram and answer the following questions.

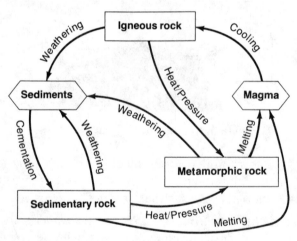

1. What processes are necessary to change a metamorphic rock into a sedimentary rock?
   (1) weathering and cementation
   (2) melting and cooling
   (3) heat and pressure

2. Which statement is true?
   (1) Sediments melt to form sedimentary rock.
   (2) Igneous rock weathers to form sediments.
   (3) Magma cools to form metamorphic rock.

3. To become igneous rock, sedimentary rock must undergo
   (1) weathering and cementation
   (2) heat and pressure
   (3) melting and cooling

Table 4-2. Common Sedimentary Rocks

Rock Name	Type of Sediment	Place of Formation
Sandstone	Sand grains	Shallow waters near a shore pounded by waves
Shale	Clay particles	Deep, calm ocean waters; lake bottoms
Limestone	Tiny seashells	Warm, shallow seas

created alter the appearance and mineral composition of the rocks, changing them into metamorphic rocks.

Marble and slate are metamorphic rocks formed from the sedimentary rocks limestone and shale. Gneiss (pronounced "nice") is a metamorphic rock that can be produced from granite, an igneous rock.

EXERCISE 1

---

1. Bedrock exposed at Earth's surface is called
   (1) a mountain   (2) a boulder   (3) an outcrop   (3) a rock fragment

2. Sandpaper is made of tiny grains of a hard mineral glued to paper and used to scrape softer substances. The best mineral to use in making sandpaper would be
   (1) talc   (2) calcite   (3) gypsum   (4) quartz

3. The graph indicates the hardness of six minerals. Which mineral is hard enough to scratch fluorite, but will not scratch garnet?
   (1) mica
   (2) calcite
   (3) hornblende
   (4) talc

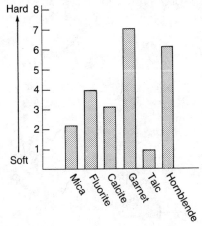

4. A *chemical* property that would help to identify a mineral is
   (1) luster   (2) hardness   (3) reaction to acid   (4) cleavage

5. Rocks that form from layers of small particles are called
   (1) metamorphic rocks   (2) sedimentary rocks   (3) igneous rocks   (4) volcanic rocks

6. *Schist* is a metamorphic rock. This means it was formed by
   (1) cooling and hardening of magma   (3) buildup of sand grains
   (2) great heat or pressure, or both   (4) buildup of clay particles

7. A mixture of small rock fragments and materials produced by living things is called
   (1) sediment   (2) gravel   (3) soil   (4) sand

8. What process was involved in forming the mountain shown below?

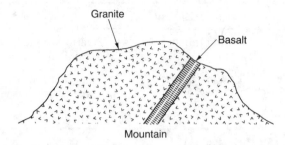

(1) cooling of magma to form igneous rocks   (3) deep underground pressure
(2) buildup of sediments in shallow water   (4) buildup of sediments in deep water

9. Granite has large mineral grains because it is formed by
   (1) slow cooling of magma   (3) rapid cooling of lava
   (2) cementation of large rock fragments   (4) high pressures

10. The illustration shows the mineral mica splitting into thin sheets. This is an example of
   (1) color
   (2) hardness
   (3) uneven fracture
   (4) cleavage

PART II. FORCES THAT CHANGE EARTH'S SURFACE

Various forces are constantly at work shaping and changing Earth's surface. Figure 4-5 illustrates these forces and their effects on Earth's surface features.

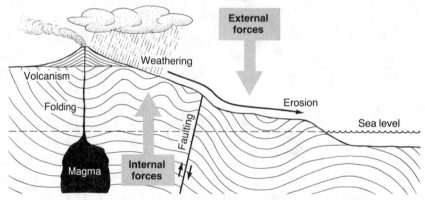

Figure 4-5. Earth's surface is shaped by the interaction of internal and external forces.

External Forces

External forces include the processes of weathering and erosion. Together, these processes wear down Earth's surface.

1. *Weathering* is the breaking down of rocks into smaller pieces. Both physical and chemical agents can cause weathering. In *physical weathering*, rock is broken into smaller fragments by physical agents. For example, when water seeps into cracks in a rock and freezes, the water expands, breaking the rock apart, as shown in Figure 4-6. The roots of plants growing in cracks can also force rocks apart.

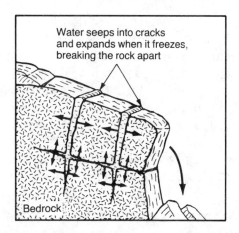

Figure 4-6.
Physical weathering caused by
water freezing in rock cracks.

Chemical weathering is the breaking down of rocks through changes in their chemical makeup. These changes take place when rocks are exposed to air or water. For instance, when rainwater combines with carbon dioxide in the air, a weak acid is formed that dissolves certain minerals in rocks and causes the rocks to fall apart. Also, when oxygen and water react chemically with iron-bearing minerals in a rock, the iron is changed into rust, which crumbles away easily.

By breaking down rocks into smaller fragments, the processes of weathering assist in the formation of soil.

2. **Erosion** is the process whereby rock material at Earth's surface is removed and carried away. Erosion requires a moving force, such as flowing water, which can carry along rock particles. This can be seen after a heavy rain, when streams turn a muddy brown from the rock material in the water.

Gravity and *water* play important roles in erosion. Gravity is the main force that moves water and rock downhill. Flowing water is very powerful; more rock material is eroded by running water than by all other forces of erosion combined. The Grand Canyon in Arizona is a spectacular example of erosion caused by running water (Figure 4-7).

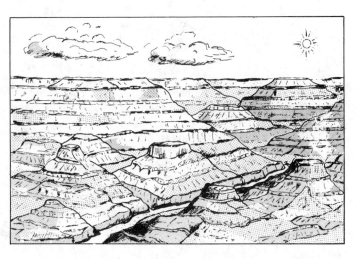

Figure 4-7.
Erosion by running water
carved the Grand Canyon,
a gorge over a mile deep.

Groundwater and glaciers are other forces that cause erosion. *Groundwater* forms from rain or snowmelt that filters into the soil. As groundwater seeps through cracks in the bedrock, the water dissolves rock material and carries it away. Eventually, this action may create large underground caves.

Glaciers are masses of ice that form in places where more snow falls in winter than melts in summer, such as in a high mountain valley. The snow that does not melt piles up over the years, and its increasing weight changes the bottom layers into ice. Gravity causes the ice to flow downhill, like a river in slow motion. As a glacier creeps along, it grinds up and removes rock material from the land surface.

Wind can also act as a force of erosion. In dry desert areas, sand grains blown along by the wind scrape and scour rock outcrops, slowly carving them into unusual shapes.

The forces of erosion are constantly at work, moving rock material from the continents into the ocean basins.

PROCESS SKILL: PREDICTING AN EXPERIMENTAL RESULT

Rocks in a stream constantly knock and scrape against each other and against the streambed as they are carried along by the flowing water. The longer the rocks are in the stream, the more they tumble about and strike one another. To simulate this action and study its effects, a student carried out the following experiment.

Twenty-five marble chips and a liter of water were placed in a large coffee can marked *A*. The can was then covered with a lid and shaken for 30 minutes. Then 25 marble chips and a liter of water were placed in a second can, marked *B*, and covered. This can was shaken for 120 minutes. The illustration below shows the materials used in the experiment. Keep in mind what you have learned about weathering to help you answer the following questions.

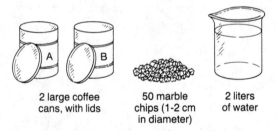

2 large coffee cans, with lids — 50 marble chips (1-2 cm in diameter) — 2 liters of water

1. Which is the best prediction of the experiment's results?
   (1) The marble chips in can *A* will be smaller and rounder than the chips in can *B*.
   (2) The marble chips in can *B* will be smaller and rounder than the chips in can *A*.
   (3) There will be no difference between the marble chips in cans *A* and *B*.

2. Which graph best predicts what would happen to rocks in a fast-moving stream over time?

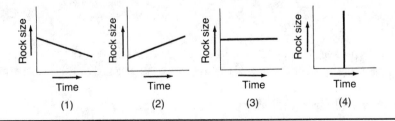

Internal Forces

Earth's internal forces also shape its surface. These forces produce *mountains*, *earthquakes*, and *volcanoes*, raising the land and building up Earth's surface.

1. **Mountains** are produced mainly by the processes of folding and faulting. **Folding** takes place when forces in Earth's crust press rocks together from the sides, bending the layers into folds. The land is squeezed into upfolds and downfolds, forming ridges and valleys (Figure 4-8).

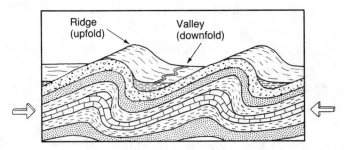

Figure 4-8. Folding: Forces in the crust can squeeze rock layers into folds.

Faulting occurs when forces in the crust squeeze or pull rock beyond its capacity to bend or stretch. The rock then breaks and slides along a crack or fracture, called a *fault*, relieving the stress in the crust (Figure 4-9). Faulting can produce mountains in a number of ways, as shown in Figure 4-10.

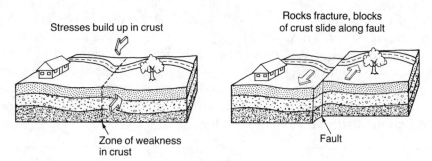

Figure 4-9. Faulting: When stresses in the crust reach the breaking point, the crust fractures and slips.

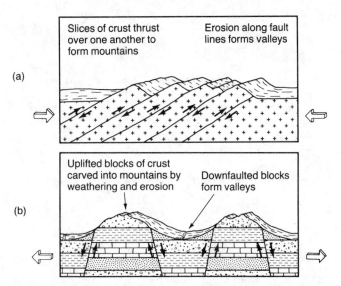

Figure 4-10. Mountains produced by (a) thrust faulting; (b) block faulting.

2. Sudden movements of rocks sliding along faults produce strong vibrations in the crust called *earthquakes*. Many earthquakes are associated with land uplift and mountain building.

3. Mountains can also be built by volcanoes. A *volcano* is a hole in Earth's crust through which lava flows from underground. During eruptions, the lava pours out onto the surface and cools to form solid rock, building upward in layers to produce a volcanic mountain, also called a volcano (see Figure 4-3 on page 43). Mount St. Helens in Washington state is a volcanic mountain.

Besides mountains, other landforms that may result from uplift include plains and plateaus.

4. *Plains* are broad, flat regions found at low elevations. They are often made of layered sedimentary rocks that were formed underwater and slowly raised above sea level.

5. *Plateaus* are large areas of horizontally layered rocks with higher elevations than plains. They can form in several ways. A large block of crust may rise up along faults to create a plateau, or a plateau may be gradually uplifted without faulting. Plateaus can also be built up by lava flows.

Plate Tectonics

There is much evidence that forces at work inside Earth have raised the level of the land. For example, many mountaintops are made of sedimentary rock that was formed originally on the ocean floor. Folds and faults seen in many rock outcrops are also signs of crustal movements caused by internal forces. Scientists explain these forces and the movements they produce by the theory of *plate tectonics*.

According to this theory, Earth's crust is broken up into a number of large pieces, or *plates*, that slowly move and interact in various ways. Some plates are spreading apart, some are sliding past each other, and some are colliding. These movements cause mountain building, volcanic activity, and earthquakes along the plate edges. Figure 4-11 shows Earth's major crustal plates.

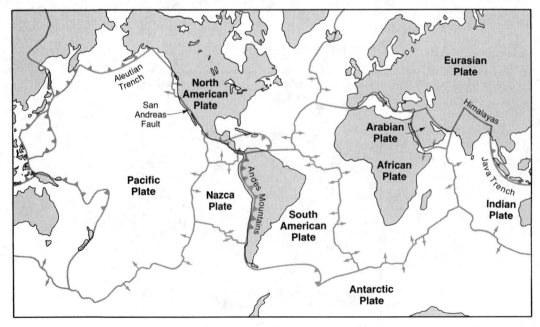

Figure 4-11. Earth's major crustal plates. (Arrows show where plates are spreading apart, triangular "teeth" show where one plate is sliding beneath another plate.)

Scientists believe that plate motions are caused by heat circulating in Earth's *mantle*, the thick zone of rock beneath the crust. The heat softens mantle rock so that it flows very slowly, following the heat currents and carrying along overlying pieces of crust (see Figure 4-12 below).

The processes of plate tectonics create many of Earth's surface features. The collision of two plates carrying continents produces great mountain ranges. The Himalayas were formed in this way.

When one plate slides sideways past another plate, a major fault and earthquake zone is produced. In California, the Pacific Plate is sliding past the North American Plate along the San Andreas fault, sometimes causing severe earthquakes.

Where plates are spreading apart, ocean basins are formed. Large continents are broken into smaller land masses that move away from each other in a process called *continental drift*. This is taking place today where the Arabian Plate is splitting away from the African Plate, opening up the Red Sea.

Ocean Floor Features

Almost three-quarters of Earth's surface is covered by ocean water. The floor of the ocean is not all flat and featureless. Scientists have found that the ocean floor has mountains, valleys, plains, and plateaus. Many of these features, such as mid-ocean ridges and ocean trenches, are produced by the processes of plate tectonics.

1. A *mid-ocean ridge* is a long, underwater mountain chain where rising magma forms new ocean crust. The new crust is added to crustal plates that spread away from the ridge, as shown in Figure 4-12. This process is called *sea-floor spreading*.

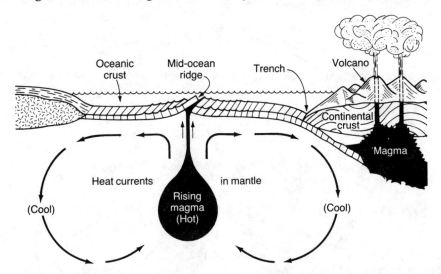

Figure 4-12. Plate tectonics: Heat currents in the mantle cause movements of Earth's crustal plates, producing many features on the seafloor and the continents.

2. *Trenches* are underwater valleys that form the deepest parts of the ocean floor. A trench is found where a plate of ocean crust collides with another plate and is forced to slide under it, back into Earth's mantle. This causes volcanic activity and mountain building along the edge of the upper plate (see Figure 4-12 above).

Other ocean floor features include continental shelves, continental slopes, the deep ocean floor, and seamounts.

3. *Continental shelves* are areas of the sea floor that slope gently away from the coastlines of most continents. The angle of slope is so slight that if you could stand on a continental shelf, you would think you were on level ground.

4. *Continental slopes* drop away from the outer edges of continental shelves to the great depths of the ocean. These slopes are much steeper than continental shelves.

Continental slopes level off into the *deep ocean floor*. The deep ocean floor is not simply a flat plain; it also has ridges and valleys. Rising here and there from the ocean floor are tall underwater mountains called *seamounts*. Most seamounts were formed by volcanoes.

When the top of a seamount rises above the water's surface, an island is formed. The Hawaiian Islands are the tops of a chain of volcanic seamounts. Figure 4-13 shows the profile of an ocean floor that includes many of these features.

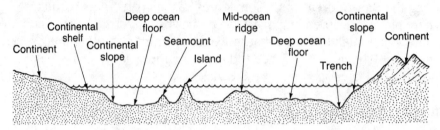

Figure 4-13. Features of the ocean floor.

EXERCISE 2

1. When water freezes in cracks in a rock, the water expands, breaking the rock apart. This is a type of
   (1) glacial erosion
   (2) physical weathering
   (3) chemical weathering
   (4) groundwater erosion

2. The diagram below shows the mineral magnetite, which contains iron, changing into rust particles. This is an example of

| Black, metallic, and magnetic | Black and rusty red, and less magnetic | Rusty red and nonmagnetic |

   (1) physical weathering
   (2) chemical weathering
   (3) erosion by running water
   (4) the role of gravity in erosion

3. Erosion is the process by which rocks at Earth's surface
   (1) are removed and carried away
   (2) crumble and decay
   (3) turn into rust
   (4) melt to form magma

4. The diagram shows how an underground cave changed over time. What process caused this?

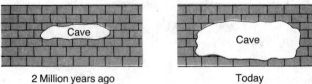

   (1) erosion by glaciers
   (2) physical weathering
   (3) erosion by groundwater
   (4) expansion of water changing to ice

5. The shape of the land in the series of diagrams below most likely changed because of

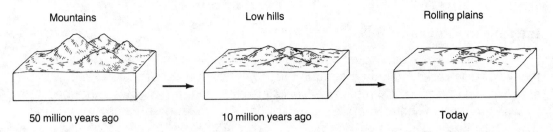

Mountains   Low hills   Rolling plains

50 million years ago   10 million years ago   Today

(1) weathering and erosion   (3) folding and faulting
(2) volcanoes and earthquakes   (4) melting

6. The rock layers in the diagram have been affected by
   (1) volcanoes
   (2) faulting
   (3) groundwater erosion
   (4) folding

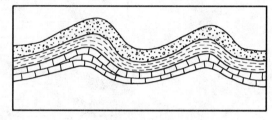

7. Mountains can be produced by all of the following processes *except*
   (1) volcanic eruptions   (2) folding   (3) weathering   (4) faulting

8. The theory that Earth's crust is broken up into large pieces that move and interact is called
   (1) evolution   (2) mountain building   (3) the rock cycle   (4) plate tectonics

9. Major mountain ranges are formed when crustal plates
   (1) push into each other   (3) move away from each other
   (2) slide past each other   (4) break into smaller plates

10. If crustal block *A*, to the left of the fault in the diagram, suddenly shifted downward several feet, what would most likely occur at location *C*?
    (1) An earthquake would occur.
    (2) A volcanic eruption would occur.
    (3) A mountain would form.
    (4) An ocean would form.

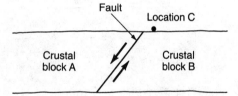

11. The ocean floor is best described as
    (1) a flat, featureless plain
    (2) having mountains, valleys, plains, and plateaus
    (3) a flat plain with a deep valley in the center
    (4) having plains and plateaus only

12. The table at the right shows the depth of the ocean at various distances from a continent. At what distance from the continent is a deep trench located?
    (1) 300 kilometers
    (2) 200 kilometers
    (3) 100 kilometers
    (4) 500 kilometers

Distance from continent	Ocean depth
50 km	400 m
100 km	9000 m
150 km	1250 m
200 km	1100 m
250 km	200 m
300 km	950 m

PART III. EARTH HISTORY

Interpreting Rocks

Scientists have pieced together much of Earth's history by studying rocks all over the world. The rocks in an area contain much information about that area's past. For example, the presence of sedimentary rocks indicates that an area was once covered by water. Fossils in sedimentary rocks tell of past life forms and the environments in which they lived.

Scientists can interpret clues in rocks that reveal the order in which they were formed. Horizontally layered sedimentary rocks are easiest to interpret. The bottom layers were laid down first and are therefore the oldest, while the youngest layers are at the top (Figure 4-14).

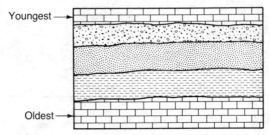

Figure 4-14. In a stack of sedimentary rocks, the oldest layers are at the bottom and the youngest at the top.

This simple situation is often complicated by later events. Folding and faulting of rock layers and invasions of igneous rock material make it difficult to tell which rocks are the oldest and which are the youngest (Figure 4-15). However, all these features are clues to events in Earth's past and the order in which they happened.

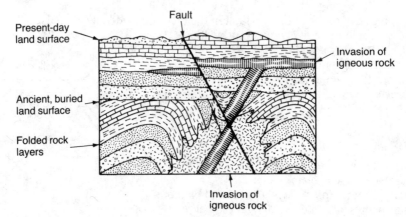

Figure 4-15. A rock sequence with a complex history of folding, erosion, volcanism, and faulting.

Fossils

Fossils are the remains or traces of organisms that lived long ago. Fossils are formed when a dead plant or animal, or some trace, like a footprint in mud, is covered by sediment that later hardens into rock. Almost all fossils are found in sedimentary rock. Figure 4-16 shows several types of fossils.

Scientists have learned much about Earth's past by studying fossils. Fossil evidence has helped scientists to trace the evolution of life from simple ancient organ-

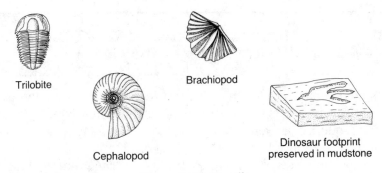

Figure 4-16. Fossils.

isms to complex present-day life forms. Fossils also provide clues to ancient environments. For example, corals live only in warm, sunlight-rich waters. Finding fossil corals in central New York State suggests that the area was once covered by a warm, shallow sea.

Fossils can sometimes be used to match up rock layers that are far apart. Finding the same group of fossils in rock layers at separate locations indicates that those layers formed in the same time period (Figure 4-17).

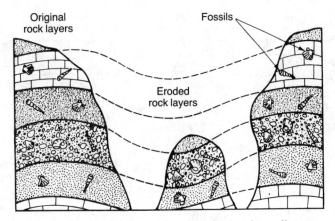

Figure 4-17. Fossils can be used to match up distant rock layers.

Dating Rocks

The relationships between rock layers, fossils, folds, faults, and invasions of igneous rock can indicate the order of events in Earth's past. However, they cannot reveal the actual age of the rocks. To determine the age of rocks, scientists use a technique called *radioactive dating*.

Most rocks contain small amounts of radioactive substances that change (decay) into nonradioactive substances at a definite rate. For example, radioactive *uranium* changes into lead at a known rate. By measuring and comparing the amounts of uranium and lead in a rock, the age of the rock can be determined.

Using this technique, scientists have been able to assign dates to major events in Earth history, such as periods of mountain building, the formation of oceans, and the appearance of various life forms. Scientists estimate that Earth itself is about four and a half billion years old.

PROCESS SKILL: EXPLAINING A RELATIONSHIP; INTERPRETING DIAGRAMS

Many events from Earth's past are recorded in rocks. By examining features in rocks, such as folds, faults, and invasions of igneous rock, the sequence of events that produced present-day rock structures can often be sorted out.

For example, observe the rock structure in Diagram 1 below. Notice that the igneous rock *cuts across* the pattern of sedimentary rock layers. For this to happen, the sedimentary layers must have already been in place. Therefore, the sedimentary rocks were formed first.

In Diagram 2, the fault has shifted the sedimentary rock layers so that they do not match up across the fault. This means that the sedimentary layers were formed first. The direction in which matching layers have been shifted indicates that the rocks on the right side of the fault have moved downward in relation to the rocks on the left side, as shown by the arrows.

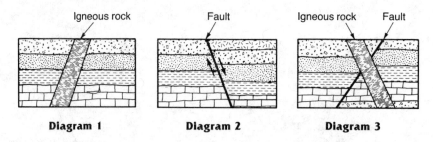

Diagram 1   Diagram 2   Diagram 3

Diagram 3 is more complex. The fault has shifted the sedimentary layers, so the sedimentary rocks were formed before the fault. The igneous rock cuts across the sedimentary layers, so the sedimentary rocks were also formed before the igneous rock. But the fault has *not* shifted the igneous rock, so the igneous rock must have formed *after*

EXERCISE 3

1. Scientists can use fossils in rocks to
   (1) match up rock layers at different places
   (2) study the evolution of life
   (3) learn about past environments
   (4) all of the above

2. Valerie identified a rock outcrop near her school as shale, a sedimentary rock. What does this suggest about the region's past?
   (1) Underground volcanic activity once took place.
   (2) Surface volcanic activity once took place.
   (3) The area was once underwater.
   (4) Great heat or pressure once affected the area.

3. By studying folds and faults in rocks, scientists can determine
   (1) the age of a rock in years
   (2) the order of past Earth events
   (3) the depth of the ocean
   (4) if there is life on Mars

the fault. The order of formation here is: sedimentary rocks, fault, igneous rock. Examine Diagram 4 and answer the questions below.

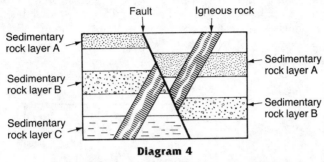

Diagram 4

1. Of the events that produced this rock structure, which occurred first?
   (1) formation of igneous rock
   (2) faulting
   (3) formation of sedimentary rock layer *C*
   (4) formation of sedimentary rock layer *A*

2. What is the correct order of formation in this rock structure?
   (1) sedimentary rocks, igneous rock, fault
   (2) sedimentary rocks, fault, igneous rock
   (3) igneous rock, sedimentary rocks, fault
   (4) igneous rock, fault, sedimentary rocks

3. Which set of arrows correctly shows the directions in which rock layers were shifted along the fault in Diagram 4?

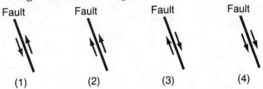

4. The diagram shows layers of sediments deposited in a body of water. Which layer was deposited first?
   (1) layer *A*
   (2) layer *B*
   (3) layer *C*
   (4) layer *D*

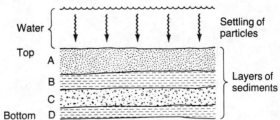

5. Most fossils are found in
   (1) igneous rocks
   (2) sedimentary rocks
   (3) metamorphic rocks
   (4) volcanic rocks

6. Carol found a coral fossil in a limestone outcrop near her home. Which statement about the area's past is most likely correct?
   (1) The area was once under deep, cold water.
   (2) The area was never underwater.
   (3) The area was once under cold, shallow water.
   (4) The area was once under warm, shallow water.

7. A rock was found containing fossils of three different ancient organisms. The table shows the time ranges in which these organisms lived. The rock must have been formed at a time when all three organisms were living. What is the best estimate of when the rock was formed?

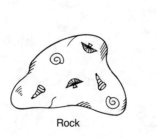

Rock

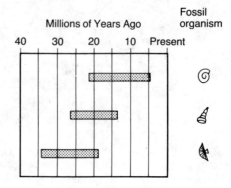

   (1) 10 million years ago   (3) 20 million years ago
   (2) 15 million years ago   (4) 25 million years ago

8. Scientists use radioactive dating to
   (1) determine the age of a rock in years
   (2) determine the order of events in Earth history
   (3) identify the rock type in a region
   (4) learn about past climate and weather

Chapter 5. Weather and Climate

PART I. WEATHER ELEMENTS AND CAUSES

Defining Weather

Surrounding Earth is a layer of gases called the *atmosphere*. These gases make up what is commonly known as *air*. **Weather** consists of the conditions of the atmosphere, such as heat, cold, sunshine, rain, snow, clouds, and wind. These conditions change from day to day and from place to place. Energy from the sun is the main cause of these changes.

Weather Elements

Weather is made up of a number of elements, including *air temperature*, *air pressure*, *humidity*, *wind speed* and *direction*, *clouds* and *cloudiness*, and *precipitation*. The weather at any given location can be described in terms of its elements.

1. Air temperature indicates the amount of heat in the atmosphere. It is measured with a *thermometer*. The two temperature scales commonly used for thermometers are the *Celsius* scale and the *Fahrenheit* scale, shown and compared in Figure 5-1. Both scales are divided into units called *degrees*.

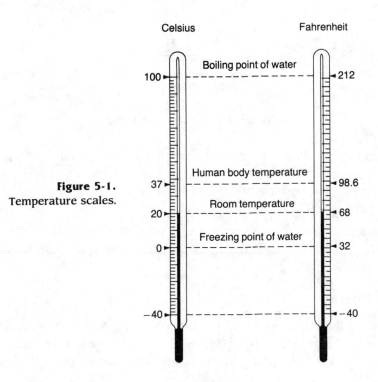

Figure 5-1. Temperature scales.

2. Because air has weight, air presses down on Earth's surface with a force called *air pressure*. This force is measured with a *barometer*, in either inches of mercury or millibars. Variations in air pressure are caused mainly by temperature and altitude. Temperature affects air pressure because cool air weighs more than warm air. Altitude affects air pressure because places at high elevations, being farther up in the atmosphere, have less air weighing down on them.

3. Humidity is the amount of *water vapor* (water in the form of a gas) present in the air. Warm air can hold more water vapor than cool air can. *Relative humidity* is the ratio between the actual amount of water vapor in the air and the maximum amount of water vapor the air can hold at that temperature. For example, a relative humidity of 90 percent means that the air contains 90 percent of the water vapor it can hold at its current temperature. Relative humidity is measured with a *hygrometer* (sometimes called a wet-and-dry-bulb thermometer).

4. Wind is the movement of air over Earth's surface. *Wind speed* is a measure of how fast the air is moving, in miles or kilometers per hour. It is determined with an *anemometer*. **Wind direction** is the direction *from which the wind is coming;* it is NOT the direction the wind is blowing *toward*. For instance, a wind blowing from north to south is called a north wind. Wind direction is determined with a *wind vane*, which points in the direction from which the wind is blowing.

Figure 5-2 shows some instruments used to measure weather elements.

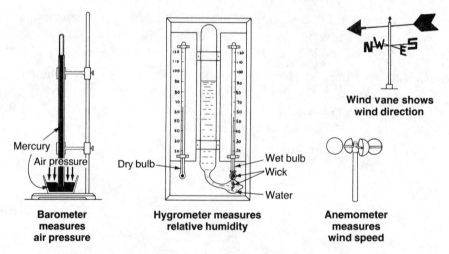

Figure 5-2. Weather instruments.

5. Clouds are masses of tiny water droplets or ice crystals, suspended in the air. *Cloudiness*, the amount of sky covered by clouds, is observed directly and described with phrases like "partly cloudy" or "mostly cloudy." A sky completely covered by clouds is described as *"overcast."*

6. Precipitation is water, in any form, falling from clouds in the sky. Precipitation can fall as rain, snow, sleet, or hail. A *rain gauge* is used to measure the amount of precipitation in inches.

The Sun's Energy

The sun is the main source of the energy in the atmosphere. As the sun heats Earth's surface, the surface radiates this heat energy back into the atmosphere. However, the sun does not heat Earth's surface evenly; consequently, the atmosphere is not heated evenly. This uneven distribution of heat energy in the atmosphere is the cause of weather.

The heating of Earth's surface depends to some extent on the nature of the surface, since some kinds of surfaces get hotter than others. For instance, pavement and sand get much hotter than do grass and water. On a larger scale, the surfaces of oceans, forests, and deserts are also affected differently by the sun. These surfaces, in turn, heat the air above them differently, producing variations in air temperature.

When air is heated, it becomes lighter (less dense) than the surrounding air. Therefore, warm air rises. Cool air is heavier (more dense), so it tends to sink. As air rises or falls, the surrounding air rushes in to replace it, causing air to circulate, as shown in Figure 5-3. This circulation, which can take place over a few kilometers or over thousands of kilometers, brings about changes in the weather.

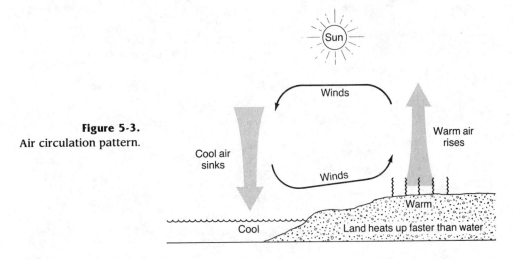

Figure 5-3.
Air circulation pattern.

The heating of Earth's surface also depends on the angle at which the sun's rays strike the surface (Figure 5-4). Near the equator, the sun's rays strike Earth most directly (Area *A* in Figure 5-4). This concentrates the sun's energy within a small area, heating the surface very effectively. However, since Earth's surface is curved, the sun's

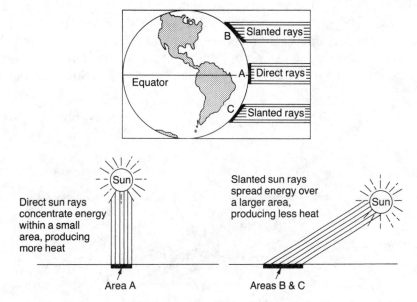

Figure 5-4. The angle of the sun's rays affects the heating of Earth's surface.

rays strike areas away from the equator at a slanting angle (Areas *B* and *C* in Figure 5-4). This spreads energy over a wider area, heating Earth's surface less effectively. The farther from the equator, the more slanted the sun's rays come in, and the less effectively they heat the surface.

The uneven heating of Earth's curved surface causes hotter air at the equator to rise and spread to the north and south, while cooler air near Earth's poles moves toward the equator to replace the rising air. Earth's rotation breaks up this simple circulation into complex global *wind belts*, shown in Figure 5-5, in which winds blow in different directions at different *latitudes*.

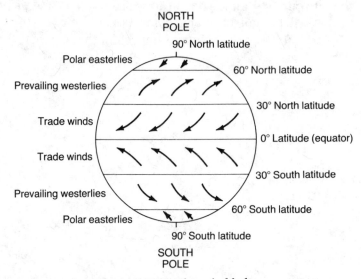

Figure 5-5. Earth's wind belts.

Latitude is distance north or south of the equator. The lines drawn parallel to the equator on a globe or a map are lines of latitude. Winds that commonly blow in the same direction at a given latitude are called the **prevailing winds**. In the United States, the prevailing winds blow from west to east, so they are called *westerlies*.

The Water Cycle

The sun's energy also powers the **water cycle**. In this process, water moves between Earth's surface and the atmosphere, as shown in Figure 5-6. Heat from the sun changes liquid water into water vapor. This is called *evaporation*. For example, evaporation takes place when a puddle of rainwater shrinks and dries up on a hot, sunny

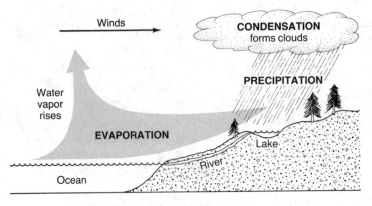

Figure 5-6. The water cycle.

day. Water enters the atmosphere by evaporation from oceans, lakes, and rivers, and by plants releasing water vapor through their leaves.

Warm air can hold more water vapor than cool air can. Therefore, when air cools, it cannot hold as much water vapor, and some of it changes back into droplets of liquid water. This process, called *condensation*, produces dew, fog, and clouds.

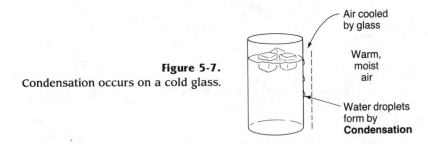

Figure 5-7.
Condensation occurs on a cold glass.

1. Dew is formed when water vapor condenses onto cool surfaces or objects. This takes place, for instance, when water droplets appear on the outside of a cold drink (see Figure 5-7).

2. Condensation may also form tiny water droplets that remain suspended in the air. When this takes place near the ground, *fog* is produced.

3. When moist air rises high into the atmosphere and cools, condensation forms *clouds*. If enough water vapor condenses, the tiny water droplets may join together into larger, heavier drops that fall to Earth as precipitation. Then the cycle of evaporation, condensation, and precipitation can begin all over again.

Climate

Climate is the general character of the weather that prevails in an area from season to season and from year to year. It can be thought of as the average weather of an area over a long period of time.

A number of factors combine to produce different climates.

1. One factor is *latitude*, which is distance from the equator. Places at high latitudes, far from the equator, tend to have colder climates than places at lower latitudes. For instance, Canada is at a higher latitude than Mexico, so it has a colder climate.

2. Another factor is *altitude*, which is the height (elevation) above sea level of a place. Higher elevations are cooler than lower elevations. Just as a mountaintop is colder than its base, a city at a high altitude will have a colder climate than a city at the same latitude that is at a lower altitude.

3. Large bodies of water can affect climate. Land areas close to oceans or large lakes generally have more moderate climates (cooler summers and warmer winters) than areas far from water. Water absorbs and gives off heat more slowly than land does. Therefore, as the land heats up during summer, the water stays relatively cool, keeping coastal areas cooler in summer than places farther inland. In winter, the situation is the opposite. The water loses heat built up during summer more slowly than the land does, keeping coastal areas warmer in winter than areas farther inland.

4. Mountain barriers can also influence climate. The side of a mountain range facing the prevailing winds tends to have a cool, moist climate, while the opposite side of the mountains has a warmer, drier climate. This is illustrated in Figure 5-8 on page 64.

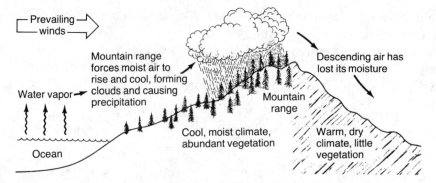

Figure 5-8. A mountain range can affect climate.

PROCESS SKILL: DESIGNING AN EXPERIMENT; PREDICTING RESULTS

Cold air is heavier and more dense than warm air. Therefore, cold air tends to sink and warm air tends to rise. To design an experiment that can demonstrate this, you would need cool and warm air, a way to control the flow of air, and several thermometers.

For instance, a small room with a window provides a means of controlling air flow (see the diagram). On a day when the temperature outside is lower than the indoor temperature by at least 10°C, you could open the window to let in cold air and measure temperature changes in the room with thermometers. If it is true that cold air sinks and warm air rises, the lower levels of the room will get colder more quickly than the higher levels will. Answer the questions that follow the diagram on this page and the top of the next page.

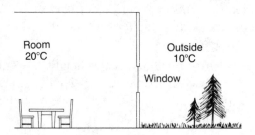

1. How should the thermometers be arranged in the room to show that cold air sinks and warm air rises?

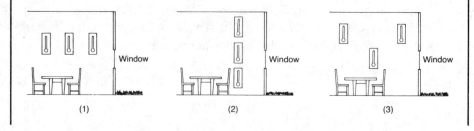

2. If this experiment were performed on a summer day, when the outdoor temperature was 10°C higher than the indoor temperature, what would probably happen in the room after opening the window?
   (1) The temperature near the ceiling would increase faster than the temperature near the floor.
   (2) The temperature near the floor would increase faster than the temperature near the ceiling.
   (3) The temperature at all levels of the room would increase at the same rate.

EXERCISE 1

1. What is the source of the energy that sets Earth's atmosphere in motion and causes weather?
   (1) volcanoes   (2) gravity   (3) the ocean   (4) the sun

2. Air would be densest and heaviest at a temperature of
   (1) 25°C   (2) 20°C   (3) 15°C   (4) 10°C

3. The average weather conditions of an area over many years determine that area's
   (1) balance of nature   (2) geologic history   (3) climate   (4) latitude

4. When it rained one morning, Willie saw a puddle of water form in the street. In the evening, the puddle was gone. The water probably disappeared by
   (1) evaporation   (2) condensation   (3) precipitation   (4) expansion

5. What temperature is indicated by the thermometer in the diagram?
   (1) 10°C
   (2) 20°C
   (3) 30°C
   (4) 25°C

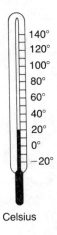

Celsius

6. In the diagram, the most likely reason for City B to be cooler than City A is that
   (1) City B is at a different latitude   (3) City B is closer to the ocean
   (2) City B is at a higher altitude   (4) City B is closer to the sun

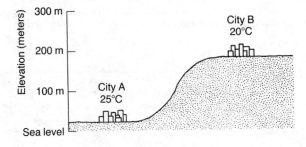

7. Mr. Wilson decided to drive his car up Pikes Peak, a tall mountain in Colorado. As he gets higher up the mountain, the air temperature will most likely
   (1) decrease
   (2) increase
   (3) remain the same
   (4) first decrease and then increase

8. On a sunny summer day, a thermometer was placed above each of the surfaces shown in the diagram. Which thermometer would most likely have had the highest temperature reading?
   (1) *A*
   (2) *B*
   (3) *C*
   (4) All would have the same reading.

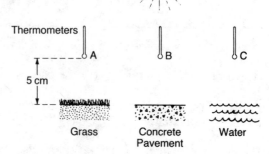

9. The diagram shows two cities, *A* and *B*, and their positions on a continent. How will the climates of the cities compare?

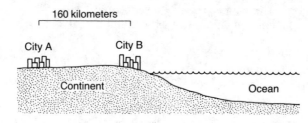

   (1) City *B* will have warmer summers and colder winters than City *A*.
   (2) City *B* will have cooler summers and warmer winters than City *A*.
   (3) City *B* will have cooler summers and colder winters than City *A*.
   (4) Both cities will have the same climate.

10. The most likely reason that New York City has a cooler climate than Miami is
    (1) a difference in distance from the ocean
    (2) a difference in altitude
    (3) a difference in air pressure
    (4) a difference in latitude

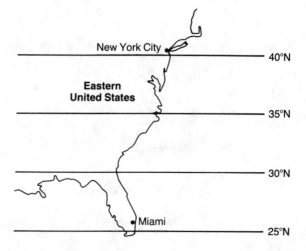

11. Which graph indicates the general temperature change as you travel from the North Pole (NP) to the South Pole (SP)?

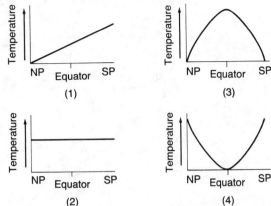

12. Which letter in the water cycle diagram indicates where condensation occurs?
    (1) *A*
    (2) *B*
    (3) *C*
    (4) *D*

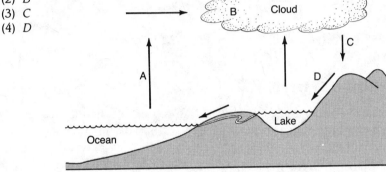

PART II. LARGE-SCALE WEATHER SYSTEMS

Air Masses

A large body of air that has uniform temperature and moisture conditions throughout it is called an ***air mass***. Much of our weather is determined by air masses.

An air mass forms when air stays over a large area of Earth's surface and takes on the temperature and moisture characteristics of that area. An air mass that forms over a warm body of water, like the Gulf of Mexico, will be warm and moist. Air masses that enter the United States from Canada are usually cold and dry because they formed over a cool land surface. An air mass builds up over an area for a few days and then begins to drift across Earth's surface.

There are four different surface conditions that affect the formation of air masses. Air masses that originate over *land* are dry. Those that form over *water* are moist. Cold air masses originate near the *poles*, and warm air masses form near the *equator*.

The major air masses that affect the continental United States, shown in Figure 5-9 (page 68), enter the country from the north, west, and south. They are then blown from west to east by the prevailing winds. As an air mass moves, it changes the local weather conditions at the surface below. The weather may become warmer or cooler, wetter or drier, depending on the type of air mass passing by.

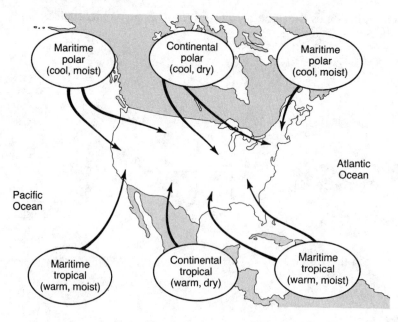

Figure 5-9. Major air masses affecting the United States.

High- and Low-Pressure Systems

Surface air pressures are usually highest in the centers of air masses, so these areas are called **high-pressure systems,** or simply *highs.* The air in a high tends to sink, and winds blow outward from the center, turning in a clockwise direction, as shown in Figure 5-10(a). High-pressure systems usually bring clear skies, dry weather, and gentle winds.

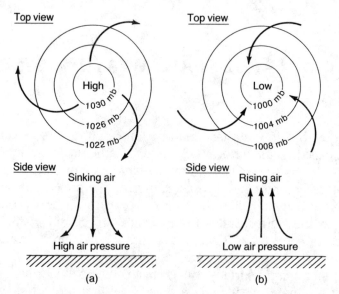

Figure 5-10. Movement of air currents in (a) a high-pressure system; (b) a low-pressure system. (mb stands for millibars.)

Surface air pressures are lower toward the edges of air masses. These areas form **low-pressure systems,** or *lows.* The air in a low tends to rise, and winds spiral in toward the center in a counterclockwise direction, as shown in Figure 5-10(b). (Note that in both highs and lows, winds always blow from areas of higher air pressure toward areas of lower air pressure.) Low-pressure systems usually bring cloudy, wet weather, often with strong, gusty winds.

Highs and lows are generally indicated on weather maps by the letters **H** and **L** (see Figure 5-13 on page 71). The highs on weather maps usually indicate the centers of air masses, while lows are generally found at the edges of air masses, or between air masses. Because highs are large and tend to move slowly, weather changes are usually gradual, rather than sudden. Changes in air pressure readings signal the passing of highs and lows.

Fronts

When one air mass comes into contact with another air mass, a boundary, called a *front*, forms between them. Sudden changes in weather conditions can occur across a front. The air masses that meet often differ in temperature, humidity, and density. These differences prevent the air masses from mixing. The cooler, drier air is heavier and remains close to the ground, while the warmer, moister air is lighter and rises upward. This causes areas of low pressure to develop along fronts, often producing clouds, strong winds, and precipitation. These lows are the major storm systems of our latitudes.

Different kinds of fronts are produced depending on how the air masses come together.

1. Cold Front. If a cold air mass pushes into and under a warm air mass, a **cold front** is formed (Figure 5-11). Cold fronts usually bring brief, heavy downpours, gusty winds, and cooler temperatures. On hot, humid summer days, the passing of a cold front typically causes thunderstorms, followed by a decrease in temperature and humidity.

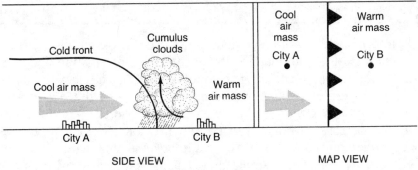

Figure 5-11. Cold front.

2. Warm Front. When a warm air mass pushes into and over a cold air mass, a **warm front** is created. The warm air slides up and over the cooler air (Figure 5-12).

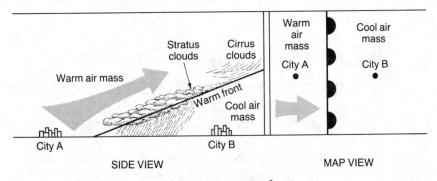

Figure 5-12. Warm front.

Warm fronts bring light precipitation lasting a day or two and warmer temperatures. When the sky is overcast all day with light rain falling, a warm front is most likely present.

Different types of clouds are produced depending upon how air rises. If air is pushed straight up, as it is along a cold front, puffy *cumulus* clouds are produced. If air rises at a low angle, as it does along a warm front, flat layers of *stratus* clouds are formed. Wispy *cirrus* clouds, which are made up of ice crystals, form high in the atmosphere. They may look like feathers, or tufts of hair. The presence of many high cirrus clouds may indicate that a warm front is approaching (see Figure 5-12 on page 69).

Weather Forecasting

Weather forecasting is an attempt to make accurate predictions of future weather. The accuracy of weather forecasting is improving as technology advances. In addition to weather balloons, thermometers, and barometers, weather forecasters now have a wide array of weather satellites, radar devices, and computer systems at their disposal.

Short-range local forecasts are comparatively easy. They are based mostly on *air pressure* readings and observations of *cloudiness* and *wind direction*. Changes in these weather elements are usually good indications of the weather for the next day or two.

(1) Decreasing air pressure readings signal the approach of stormy weather, while rising air pressure suggests that fair weather is coming. (2) An increase in cloudiness

PROCESS SKILL: ORGANIZING INFORMATION INTO A TABLE

Information can often be put into an orderly format by creating a chart or table that summarizes it. For instance, information about air masses can be organized in the form of a table. Air masses are named for the temperature and moisture characteristics of the region over which they form. *Continental* air masses form over land and are dry. *Maritime* air masses form over water and are moist. *Tropical* air masses form in the south and are warm, while *polar* air masses form in the north and are cool. For example, an air mass that formed over Canada would be called *continental polar*, indicating that it is cool and dry.

Types of Air Masses

Name of Air Mass	Temperature (Warm or Cool)	Humidity (Moist or Dry)
Continental Polar	Cool	Dry
Maritime Polar	?	Moist
?	Warm	Dry
Maritime Tropical	?	?

The chart above represents some of this information. Copy the chart and fill in the missing information.

Based on your completed table, how would you describe a *continental tropical* air mass in terms of its temperature and moisture content?

is a sign that a front is approaching, probably bringing precipitation. (3) In New York State, winds blowing from the west usually bring fair weather, while winds from the south or east often bring wet weather.

Today, weather forecasters also use information from weather satellites and radar to improve their short-range forecasts. This information is used to produce up-to-date weather maps like the one in Figure 5-13. Such maps can help us predict coming

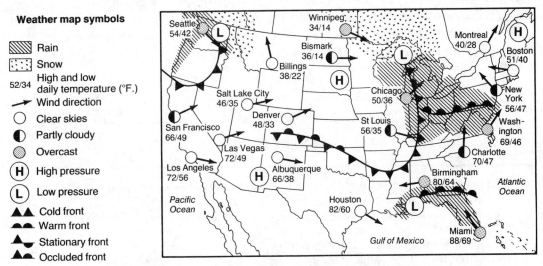

Figure 5-13. A weather map is useful for making predictions.

The following incomplete table contains information about air pressure systems. Copy the table and refer to the section in the text about highs and lows to help you fill in the missing information. Then answer the questions below.

Air Pressure Systems

| Type of Air Pressure System | Characteristics | | | |
	Vertical Air Movement	Horizontal Air Movement	Type of Weather	Location in Air Mass
High	Sinking	?	Fair	?
Low	?	In toward center, turning counterclockwise	?	Edges

1. The movement of air in a high-pressure system is
   (1) counterclockwise and rising
   (2) counterclockwise and sinking
   (3) clockwise and rising
   (4) clockwise and sinking

2. The weather on the edge of an air mass is usually
   (1) stormy, with low air pressure
   (2) stormy, with high air pressure
   (3) fair, with low air pressure
   (4) fair, with high air pressure

changes. Weather systems generally move from west to east across the United States. Therefore, if a weather map shows a high immediately to our west, we can forecast fair weather for the next day. Conversely, if the map shows a low to our west, we can expect stormy weather.

Long-range weather forecasting is more difficult. Computers, satellite photographs, and radar images enable forecasters to track the movements of large-scale weather systems like air masses and fronts. With this information, they can make predictions of weather several days in the future. However, such forecasts are not always accurate. Generally, weather forecasts for the next day or two are reliable, while long-range forecasts usually have to be revised.

EXERCISE 2

1. During winter, air masses that form over northern Canada often affect the weather in New York State. Such an air mass would be
   (1) dry and warm   (2) dry and cool   (3) moist and warm   (4) moist and cool

2. Air pressure within an air mass is usually
   (1) high in the center and low on the edges
   (2) high on the edges and low in the center
   (3) constant throughout
   (4) varying throughout

3. The major storm systems of our latitudes are
   (1) high-pressure systems
   (2) cold, dry air masses
   (3) tropical air masses
   (4) low-pressure systems

4. On a weather map, Mark saw the word "High" over New York State. This indicates
   (1) high temperature   (2) high clouds   (3) high air pressure   (4) high winds

5. At which city in the diagram would the air pressure most likely be the greatest?
   (1) City *A*
   (2) City *B*
   (3) City *C*
   (4) City *D*

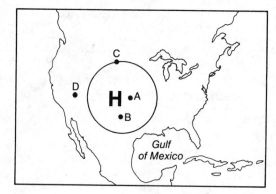

6. Changes in air pressure indicate
   (1) the change of seasons
   (2) the passing of highs and lows
   (3) the climate is changing
   (4) the sun is setting

7. Jimmy listened to the weather report on television. The forecaster said a warm front was approaching. Jimmy knew he could expect
   (1) fair weather
   (2) brief thunderstorms
   (3) tornadoes
   (4) light rain for about a day or two

8. Margaret observed the weather from Friday night to Saturday night. First it was hot and humid, then there were thunderstorms, and finally the air became cooler and drier. These changes were probably due to the passing of a
(1) warm front   (2) cold front   (3) hurricane   (4) wind belt

9. The diagram shows an air mass entering the United States from the northwest. This air mass formed over the North Pacific Ocean, so it would be
(1) moist and warm
(2) moist and cool
(3) dry and warm
(4) dry and cool

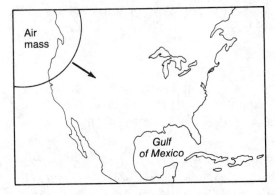

10. In the diagram, the air mass most likely to affect New York State the next day would be
(1) air mass A
(2) air mass B
(3) air mass C
(4) air mass D

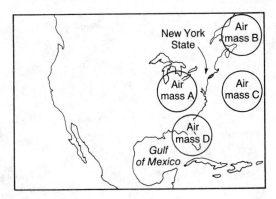

Questions 11 and 12 refer to the following diagram.

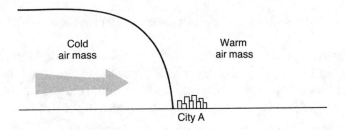

11. The line separating the cold air mass from the warm air mass represents a
(1) warm front   (2) cold front   (3) line of latitude   (4) high-pressure system

12. City A is most likely about to experience
(1) a light rain, followed by warmer temperatures
(2) heavy rains lasting several days
(3) no change in weather conditions
(4) brief downpours, followed by cooler temperatures

PART III. WEATHER HAZARDS AND POLLUTION

Weather Hazards

Weather directly affects our lives. Lack of rain can ruin crops and force emergency measures such as water rationing. Too much rain can cause destructive floods. Storms produce severe weather conditions that can make travel dangerous, force schools and businesses to close, and cause deaths, injuries, and property damage.

Storms

Storms are natural disturbances in the atmosphere that involve low air pressure, clouds, precipitation, and strong winds. The major types of storms are *thunderstorms*, *hurricanes*, *tornadoes*, and *winter storms*. Each type has unique characteristics and dangers.

1. Thunderstorms are brief, intense storms that affect a small area. They are produced when rapidly rising warm air causes cumulus clouds to build upward into a *thunderhead* (Figure 5-14). Thunderstorms are characterized by thunder and lightning, strong gusty winds, and sometimes hail.

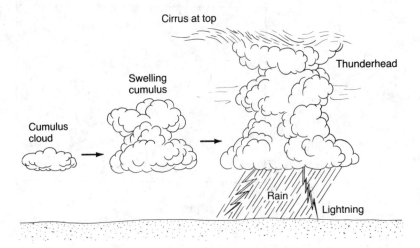

Figure 5-14. Development of a thunderhead.

Lightning is a huge electrical discharge, like a giant spark. Lightning strikes are very dangerous and can be fatal. Large hailstones that sometimes fall during thunderstorms can also be dangerous. During a thunderstorm, you should stay indoors and especially avoid hilltops, open fields, beaches, and bodies of water.

2. Hurricanes are huge, rotating storms that form over the ocean near the equator. They produce very strong winds, heavy rains, and large, powerful waves. A calm region in the storm's center is called the *eye* of the hurricane.

Hurricanes can cause severe flooding and damage from their high winds. People living along the coast and in flood-prone regions should leave their homes and move to higher ground when a hurricane strikes.

3. Tornadoes are violently whirling winds, sometimes visible as a funnel-shaped cloud (Figure 5-15). They are produced by particularly severe thunderstorms. Tornadoes usually appear suddenly, carve a narrow path of destruction, and disappear as suddenly as they came. Spiraling high-speed winds and extremely low air pressure are the unique features of tornadoes.

Figure 5-15.
Tornado funnel cloud.

A tornado can lift and toss large objects, including cars, into the air. It can destroy houses in its path in a matter of seconds. An underground cellar or basement is the safest place to be during a tornado.

4. Winter storms include blizzards and ice storms. *Blizzards* are fierce storms with strong winds, blowing snow, and very cold temperatures. *Ice storms* occur when falling rain freezes at Earth's surface, coating everything with ice. Under these conditions, you should remain indoors and not attempt to travel.

Pollution

Human activities can affect the atmosphere and the weather. Factories, power plants, cars, and airplanes produce harmful substances called **pollutants**. The build-up of pollutants in the atmosphere can cause a number of weather problems.

Smog is a haze formed by the reaction of sunlight with chemicals in automobile exhaust and factory smoke. Smog tends to hang over large cities, giving the air a hazy, dirty look. Inhaling smog is very dangerous for people with breathing problems like asthma, and harmful to the lungs of even healthy people.

Chemicals in smoke from factories and vehicles can also increase the *acidity* of the moisture in clouds. When this moisture falls to Earth as **acid rain**, it can harm lakes and forests, and the creatures that live in them. Because air pollutants are often carried along by the prevailing winds, acid rain may fall far from the source of pollution (Figure 5-16).

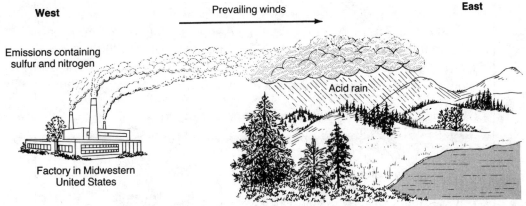

Figure 5-16. Acid rain may affect areas far downwind from the source of pollution.

Many industries and most forms of transportation produce *carbon dioxide*. In the atmosphere, carbon dioxide acts like the glass of a greenhouse, trapping heat close to Earth instead of letting it radiate back into space. Many scientists fear that the buildup of carbon dioxide in the atmosphere caused by human activities may lead to *global warming*, a rise in worldwide average temperatures. This condition, often called

PROCESS SKILL: INTERPRETING DATA IN A TABLE

Storms form when the temperature, air pressure, and moisture conditions necessary for their development exist. At certain times of the year, weather conditions are more likely to produce one of the three types of storms—thunderstorms, tornadoes, and hurricanes—than at other times. Therefore, each storm type should have a "storm season."

You can determine if there are "storm seasons" by keeping track of the storms of each type that strike during the year, and recording which months they occur in, as shown in the table below. With this information, you can see if each kind of storm takes place mostly within a definite time of year—a "storm season."

Number of Storms Per Month

| | Type of Storm | | |
Month	Thunderstorm	Tornado	Hurricane
January	0	1	0
February	1	0	0
March	3	2	0
April	6	4	0
May	9	5	0
June	15	4	0
July	18	2	1
August	16	1	5
September	10	1	8
October	6	0	4
November	2	1	1
December	0	1	0

According to the table, most thunderstorms occur in the months of June, July, and August, so there does appear to be a thunderstorm season. Is there evidence in the table of a season for tornadoes? How about for hurricanes? Use the table to answer the following questions.

1. During which three months do most tornadoes occur?
   (1) August, September, and October
   (2) April, May, and June
   (3) June, July, and August
   (4) May, June, and July

2. Which storm type has the most clearly defined "storm season"?
   (1) thunderstorm   (2) tornado   (3) hurricane
   (4) all three types have equally well-defined seasons

the *greenhouse effect,* could have disastrous results, making climates hotter and drier, and interfering with agriculture. Polar ice caps could melt, raising the sea level and flooding coastal areas and many major cities.

Certain natural events also release pollutants. Forest fires and volcanic eruptions give off huge quantities of dust and ash particles that collect high in the atmosphere, blocking sunlight and causing cooler temperatures on Earth. Plants release irritating pollen into the air, causing health problems for people with hay fever and asthma. There is little that we can do to control such natural pollutants. However, we can, and must, control our own activities that cause pollution if we are to avoid harming our planet.

EXERCISE 3

---

1. The main hazard of a thunderstorm is
   (1) thunder   (2) funnel-shaped winds   (3) heavy rain   (4) lightning

2. The violent windstorm visible as a funnel-shaped cloud in the illustration is called a
   (1) hurricane
   (2) thunderstorm
   (3) tornado
   (4) blizzard

3. Substances released into the atmosphere by the activities of humans
   (1) always have a positive effect on the weather
   (2) can have a harmful effect on the weather
   (3) have no effect on weather
   (4) may cause global cooling

4. John visited the city on a warm summer day. He noticed that the air was hazy, and that his eyes and throat burned. This was probably caused by
   (1) an approaching storm   (2) smog   (3) low clouds   (4) global warming

5. When scientists speak of the "greenhouse effect," they are referring to
   (1) the fact that vegetables grown in a greenhouse do not taste as good as vegetables grown outdoors
   (2) the use of green paint on houses to camouflage them
   (3) the idea that pollution caused by human activities may lead to global warming
   (4) the fact that ash from volcanic eruptions can cause cooler temperatures

6. A hurricane is approaching the east coast of Florida. What dangers should the people there take precautions against?
   (1) cold temperatures, blowing snow, and poor visibility
   (2) funnel-shaped winds that can lift large objects
   (3) lightning and hailstones
   (4) flooding, large waves, and strong winds

7. The map shows a major industrial city and three lakes in the central U.S. Which lake is most likely to be affected by acid rain caused by pollution from the city?

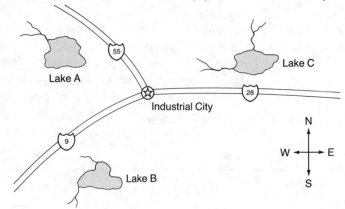

Lake A

55

Lake C

28

Industrial City

9

Lake B

N
W ← → E
S

(1) lake *A*   (2) lake *B*   (3) lake *C*   (4) all three lakes will be affected equally

8. The table below lists sources of air pollutants and the percentage contributed by each source.

Sources and Percentages of Air Pollutants

Pollutant Source	Percentage of Total Pollutants
Transportation	42%
Fuel	21%
Solid waste disposal	5%
Forest fires	8%
Miscellaneous	10%
Industrial processes	14%

Which pie graph correctly represents the data in the table?

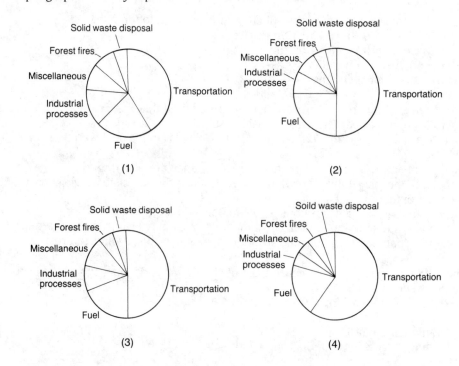

(1)

(2)

(3)

(4)

Chapter 6. Astronomy and Aerospace Science

PART I. EARTH'S MOTIONS IN SPACE

Earth's Shape

The planet Earth is a solid body shaped like a ball, or *sphere.* It is composed mostly of rock, with an atmosphere of gases surrounding it and liquid water covering nearly three-fourths of its surface. Figure 6-1 shows the size, shape, and surface structure of Earth.

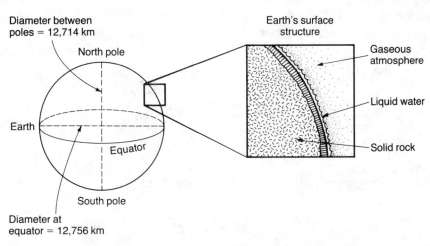

Figure 6-1. Earth's shape, size, and surface structure.

Earth's Rotation

People once believed that Earth stands still while the sun, moon, stars, and planets revolve around Earth each day. This seems reasonable, since we do not feel Earth moving, but the sun does appear to move across the sky during the day, and the moon and stars appear to move at night. However, today we know that these *apparent* motions of the sun, moon, and stars are actually caused by Earth's own motion.

Earth spins like a top. This spinning motion is called *rotation*. Earth rotates from west to east—or put another way, looking down at the North Pole from space, we see Earth spinning in a counterclockwise direction.

Extending through Earth between the North and South Poles is an imaginary rod, or *axis of rotation*, around which Earth spins (see Figure 6-2 on page 80). A basketball spinning on a fingertip gives a good idea of how Earth spins on its axis. The line from the fingertip through the basketball to the top of the ball is the axis of rotation.

The rotation of Earth has several results. (1) In particular, Earth's rotation causes the daily change from day to night. At any given time, half of Earth is in daylight,

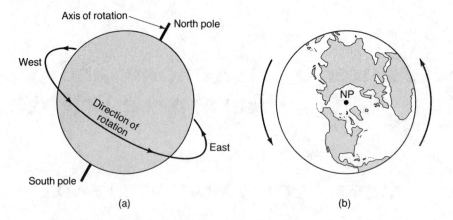

Figure 6-2. Rotation of Earth: (a) Earth rotates from west to east around its axis. (b) Viewed from above the North Pole (NP), Earth rotates in a counterclockwise direction.

facing the sun, while half is in darkness, facing away from the sun. This is shown in Figure 6-3. Every day, all places on Earth—except the areas near the poles—experience this change from daylight to darkness. (Areas within the Arctic and Antarctic Circles experience several weeks of continuous daylight or darkness at certain times of the year.)

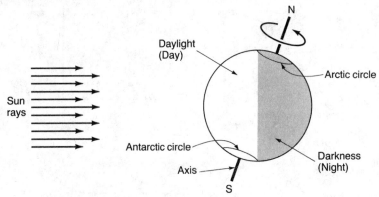

Figure 6-3. Earth's rotation causes the change from day to night.

(2) The speed of Earth's rotation causes the length of one day to be about 24 hours. This is the amount of time Earth takes to rotate once on its axis.

(3) The daily motions of the sun, moon, planets, and stars across the sky are also caused by Earth's rotation. These objects appear to rise in the eastern sky and set in the western sky because Earth rotates from west to east.

Proof of Earth's Rotation

Before the era of spaceflight, Earth's rotation could not be directly observed. However, in the 1850s, a French scientist named Foucalt was able to prove that Earth rotates. He pointed out that a freely swinging pendulum appears to change its direction of swing over time (Figure 6-4). But according to the laws of physics, a pendulum will not change direction if it is not interfered with. Therefore, Foucalt reasoned that the pendulum's apparent change in direction must be caused by Earth rotating beneath the swinging pendulum.

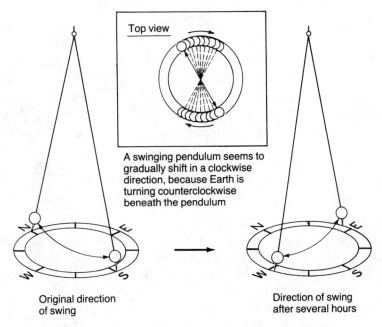

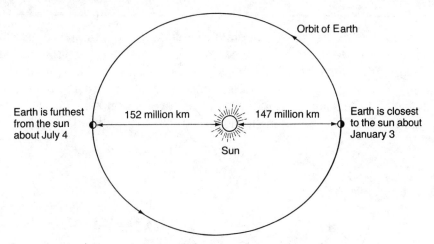

Figure 6-4. A pendulum's direction of swing seems to gradually shift, because Earth is rotating beneath it.

Earth's Revolution

Earth moves around the sun in a motion called ***revolution***. The path Earth travels around the sun is called an ***orbit***. Earth's orbit, although nearly circular, is actually slightly oval in shape (Figure 6-5).

Figure 6-5. Earth's orbit around the sun is an oval, not a perfect circle.

Earth's revolution has two major effects. First, the time Earth takes to revolve once around the sun defines the length of a year. During that time, Earth rotates on its axis 365 ¼ times, so there are 365 ¼ days in a year. For convenience, the calendar year is 365 days, and an extra day is added every fourth year (called a *leap year*) to make up for each leftover ¼ day.

Second, Earth's revolution around the sun, combined with the *tilt of Earth's axis*, causes the changing seasons on Earth. Earth's axis of rotation is not *perpendicular* (at a right angle) to the plane of its orbit; rather, it is tilted (Figure 6-6, page 82).

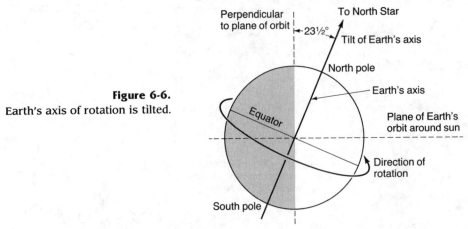

Figure 6-6.
Earth's axis of rotation is tilted.

No matter where Earth is in its orbit, the axis is always tilted in the same direction in space, pointing toward the North Star. While all the other stars seem to move across the night sky, the North Star remains motionless because Earth's axis points to it.

Seasonal Changes

Earth's orbit takes it closest to the sun in early January and farthest from the sun in early July (see Figure 6-5 on page 81). This means it is *not* Earth's changing distance from the sun that causes the changing seasons. The real cause is the tilt of Earth's axis as Earth revolves around the sun. Because the axis always points in the same

---

PROCESS SKILL: MAKING QUANTITATIVE OBSERVATIONS; PERFORMING CALCULATIONS

Earth's axis of rotation points to Polaris, the North Star. As Earth rotates, the stars closest to Polaris appear to circle it. This is called *circumpolar star motion*. Although this motion is very slow, it can be detected using time-lapse photography. By pointing a camera toward Polaris and leaving the shutter open for several hours, a star trail picture like the one shown in Diagram 1(a) can be produced.

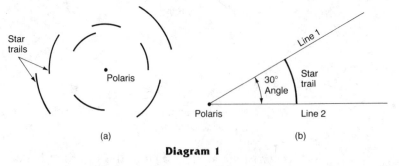

Diagram 1

Every 24 hours, each star makes a complete circle around Polaris. By dividing this number of hours into the number of degrees in a circle (360), we find that in one hour a star moves 15° around the circle. To measure the angle produced by a star trail, draw a line from each end of the star trail to Polaris, as shown in Diagram 1(b). The angle formed by the intersecting lines can be measured with a protractor.

How long did it take to produce the star trail in Diagram 1(b)? The

direction while Earth orbits the sun, the Northern Hemisphere is tilted toward the sun part of the year and away from the sun part of the year (Figure 6-7). This causes changes in the length of daylight and in the angle at which the sun's rays strike Earth.

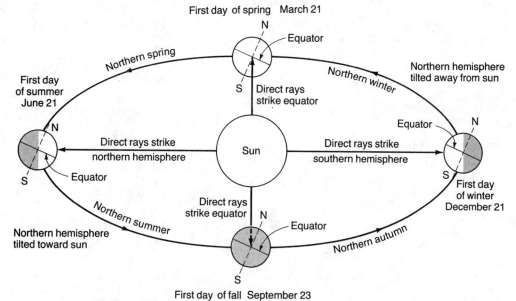

Figure 6-7. The seasons are caused by the tilt of Earth's axis and Earth's revolution around the sun.

star trail angle is 30°, so the star has traveled 30° around a circle. We know that a star moves 15° around a circle in one hour. Dividing 30 by 15, we get 2 hours as the amount of time it took to produce the star trail. Base your answers to the following questions on Diagram 2 below, which shows circumpolar star trails.

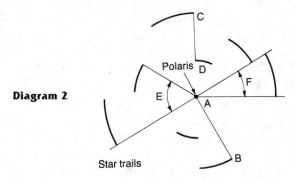

Diagram 2

Star trails

1. Which measurement should be taken to determine the length of time of the motion?
   (1) line *AB*   (2) line *CD*   (3) angle *E*   (4) angle *F*

2. If angle *F* is 23°, the amount of time taken to produce these star trails was about
   (1) a half hour   (3) one and a half hours
   (2) one hour   (4) two and a half hours

In summer, the Northern Hemisphere is tilted toward the sun and the sun follows a higher path across the sky. The sun's rays strike Earth more directly and daylight lasts longer (Figure 6-8). Areas north of the equator are therefore heated most effectively, producing the hot season. On the first day of summer in the Northern Hemisphere, the sun follows its highest path across the sky, and we have our longest period of daylight and shortest period of darkness (night).

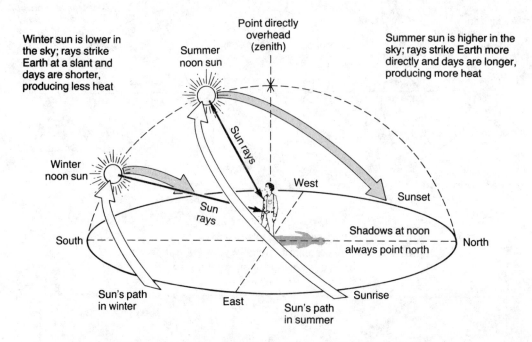

Figure 6-8. The sun's path across the sky changes with the seasons.

In winter, the Northern Hemisphere is tilted away from the sun, so the sun follows a lower path across the sky. Its rays come in at a low angle and the days are shorter. Areas north of the equator are heated less effectively and we experience our cold season. On the first day of winter, the sun travels its lowest path across the sky, and we have our shortest day and longest night.

In the Southern Hemisphere, the situation is reversed. During summer in the Northern Hemisphere, the Southern Hemisphere experiences winter. During winter in the Northern Hemisphere, the Southern Hemisphere has its summer.

During spring and autumn (fall), neither hemisphere is really tilted toward the sun. Both hemispheres experience moderate temperatures, between the extremes of summer and winter. On the first day of spring, and again on the first day of fall, the sun's rays shine directly on the equator, and day and night are of equal length. Table 6-1 on the facing page summarizes information about seasonal dates in the Northern Hemisphere.

During any season, the sun is highest in the sky each day at noon. In the continental United States, the noon sun never reaches the point directly overhead, but is always in the southern half of the sky (see Figure 6-8 above). As a result, shadows at noon always point in a northerly direction.

Proof of Earth's Revolution

Proof that Earth revolves around the sun comes from observations of stars. Stars in the night sky form patterns that have reminded people of animals or characters in

Table 6-1. Seasonal Information for the Northern Hemisphere

Date	Season	Sun's path across sky	Length of daylight and darkness
June 20, 21	First day of summer	Sun follows highest path across sky	Longest period of daylight; shortest period of darkness
September 22, 23	First day of fall	Sun follows path midway between summer and winter extremes	Daylight and darkness of equal length (12 hours each)
December 21, 22	First day of winter	Sun follows lowest path across sky	Shortest period of daylight; longest period of darkness
March 20, 21	First day of spring	Sun follows path midway between summer and winter extremes	Daylight and darkness of equal length (12 hours each)

myths. These patterns are called *constellations*. Two easily recognized constellations are Ursa Major—the Great Bear (which contains the Big Dipper) and Orion—the Hunter.

During the course of the year, different constellations become visible at night. This suggests that Earth's night side faces different directions in space (Figure 6-9), so Earth's position in relation to the sun must be changing. In other words, Earth must be moving around the sun.

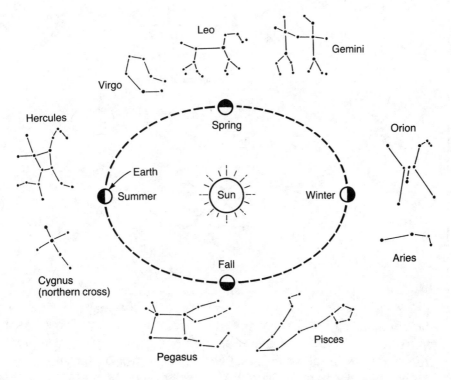

Figure 6-9. As Earth orbits the sun, different constellations become visible at night.

PROCESS SKILL: DESIGNING A MEASUREMENT PROCEDURE

As you have learned, in the United States the noon sun is never directly overhead, but is always in the southern sky. However, the noon sun is much higher overhead in summer than in winter. In fact, the sun's position at noon changes a little each day, moving higher in the sky from December 21 to June 21, and lower in the sky from June 21 to December 21. As the height of the sun changes, so does the angle of its rays striking Earth. At any location on Earth, the angle at which the sun's rays strike Earth's surface at noon changes from day to day.

You can determine the changing angle of the sun's rays by measuring the shadow of a vertical pole. Because the noon sun is always in the southern sky, the shadow of the pole is always cast to the north. The length of the shadow each day indicates the changing height of the noon sun. As the noon sun moves higher in the sky, the angle of its rays becomes larger and the length of the shadow gets shorter. As the sun moves lower, the angle of its rays becomes smaller and the shadow gets longer. This is shown in Diagram 1.

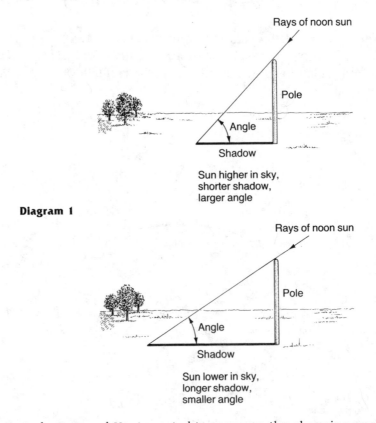

Diagram 1

A student named Kent wanted to measure the changing angle of the noon sun's rays near Albany, New York. He set up an experiment with a vertical pole and a tape measure running north along the ground from the base of the pole, as shown in Diagram 2 on the next page. For eight weeks, Kent measured the length of the pole's shadow every

Friday at 12:00 noon. Study the diagrams and then answer the following questions.

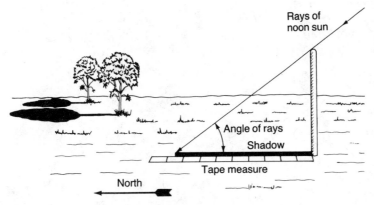

Diagram 2

1. If the length of the shadow increased each week during the experiment, this would indicate that the noon sun was
   (1) getting higher in the sky, producing a larger angle
   (2) getting higher in the sky, producing a smaller angle
   (3) getting lower in the sky, producing a larger angle
   (4) getting lower in the sky, producing a smaller angle

2. If Kent had started the experiment on June 21 and obtained a shadow length of 51 centimeters, the length of the shadow on July 15 would probably be
   (1) longer than 51 cm   (2) shorter than 51 cm   (3) exactly 51 cm

EXERCISE 1

1. The sun rises in the east and sets in the west because Earth rotates from
   (1) west to east   (2) east to west   (3) north to south   (4) north to west

2. The daily change from daylight to darkness is caused by
   (1) Earth's revolution around the sun   (3) Earth's rotation on its axis
   (2) the tilt of Earth's axis   (4) the sun's light going out at night

3. Earth rotates once on its axis in one
   (1) year   (2) week   (3) month   (4) day

4. The diagram below shows Earth as seen from above, looking down at the North Pole (NP). At location *A*, the time is 12:00 noon. What is the time at location *C*?

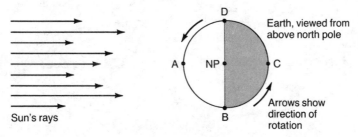

   (1) 3:00 P.M.   (2) 6:00 P.M.   (3) 12:00 midnight   (4) 6:00 A.M.

5. Jason saw the moon over a tree at 9:00 P.M. (position *A* in the diagram). An hour later the moon had moved to position *B*. This change in position was caused by the
   (1) changing of Earth's tilt
   (2) rotation of Earth
   (3) revolution of Earth
   (4) rotation of the moon

6. Earth makes one revolution around the sun in one
   (1) year (2) week (3) month (4) day

7. Tracy read in the newspaper that there would be exactly 12 hours of daylight and 12 hours of darkness the next day. The next day's date could be
   (1) October 21 (2) June 21 (3) December 21 (4) March 21

8. On July 10, the most direct rays of the sun strike Earth
   (1) in the Northern Hemisphere   (3) at the equator
   (2) in the Southern Hemisphere   (4) at the North Pole

9. At noon one day, Bob noticed the shadow of a tree in Newburgh, New York. What direction is Bob facing?
   (1) north
   (2) east
   (3) south
   (4) west

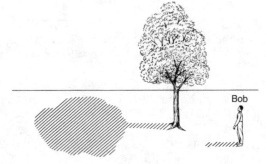

10. The diagram shows Earth at four positions in its orbit. At which position would the Northern Hemisphere be experiencing winter?

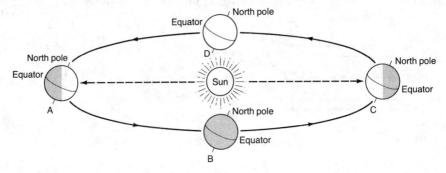

   (1) *A* (2) *B* (3) *C* (4) *D*

11. Which constellation would be visible in the night sky from the position of Earth shown in the diagram?

(1) Scorpio (2) Sagittarius (3) Aquarius (4) Gemini

12. The following question refers to the table below.

Length of Daylight for Three Cities in the U.S.

	Atlanta	New York	Boston
Latitude:	25°N	40°N	45°N
Date	Hr:Min	Hr:Min	Hr:Min
May 1	13:35	13:53	14:15
June 1	14:21	14:49	15:02
July 1	14:23	14:58	15:14

In the month of June, as you travel north from Atlanta to Boston, the length of daylight
(1) increases
(2) decreases
(3) remains the same
(4) first increases, then decreases

PART II. THE MOON

The moon is a ball of rock that revolves around Earth as Earth revolves around the sun. It is Earth's only natural satellite, and our nearest neighbor in space. There is no water or air on the moon, so it cannot support life. Because the moon is much smaller than Earth, the moon's gravity is too weak to hold moisture or an atmosphere at its surface.

The moon has a variety of surface features. The dark areas of the moon are low, flat plains. The light areas are mountainous highlands. The moon's surface is pock-marked by numerous *craters*, which are circular pits ringed by walls, or rims. Most of the craters were formed by rock fragments striking the moon's surface.

Motions and Phases of the Moon

The moon takes 29½ days, about the length of one month, to revolve around Earth from one full moon to the next (see Figure 6-10 below). As the moon orbits Earth, it also rotates on its axis. Because the moon completes one rotation in the same amount of time it takes to orbit Earth once, the same side of the moon always faces Earth. Not until spacecraft circled the moon taking photographs did we learn what the moon's far side looks like.

The moon does not give off its own light, but reflects light from the sun. The sun lights half of the moon's surface at all times, just as it always lights half of Earth. As the moon revolves around Earth, we see varying amounts of its lighted side, so the shape of the moon appears to change. These apparent changes in shape are called the *phases* of the moon.

At the phase called *new moon*, the moon is between Earth and the sun. The side facing Earth is dark, so the moon cannot be seen. At *full moon* phase, the moon is on the opposite side of Earth from the sun, and the side facing Earth is completely illuminated. Figure 6-10 shows the major phases and their relationship to the moon's revolution around Earth. The moon takes about 3⅔ days to move between each phase.

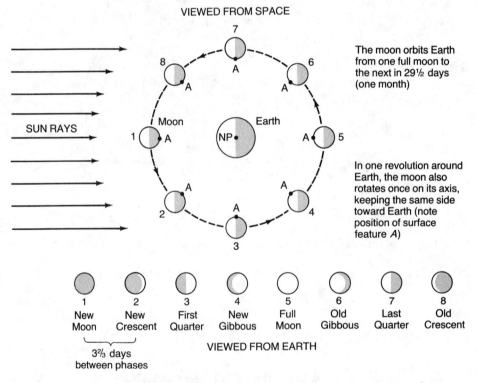

Figure 6-10. Phases of the moon: The moon takes 29½ days to complete the cycle of its phases.

Eclipses

Events called *eclipses* sometimes occur when the sun, Earth, and moon are lined up. A *lunar eclipse* takes place when the moon passes through Earth's shadow. This can happen only when the moon is full, as shown in Figure 6-11(a). A lunar eclipse lasts a few hours.

A *solar eclipse* takes place when the moon's shadow is cast on Earth, which can occur only during new moon phase, as shown in Figure 6-11(b). A solar eclipse is

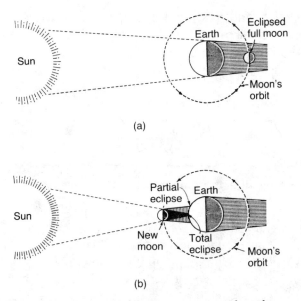

(a)

(b)

Figure 6-11. Eclipses: (a) a lunar eclipse occurs when the moon passes into Earth's shadow; (b) a solar eclipse occurs when the moon's shadow is cast upon Earth.

more brief than a lunar eclipse. A *total eclipse*, in which the sun is completely blocked by the moon, lasts only for a few minutes.

The Moon and Earth's Tides

The *tides* are the rise and fall in the level of ocean waters that take place twice each day. These changes in sea level are caused by the gravitational pull of the moon, and, to a lesser extent, the sun. The moon is much closer to Earth than the sun is, so the moon's gravity affects Earth's tides more strongly than does the sun's gravity.

The pull of the moon's gravity draws the ocean waters into two large bulges, one on the side of Earth facing the moon and one on Earth's opposite side. This causes high tides at those positions (Figure 6-12). Halfway between the tidal bulges the ocean level falls, producing low tides. As Earth rotates, locations on Earth experience changing tides. The gradual change from high tide to low tide takes about 6 ¼ hours, and another 6 ¼ hours for the water to rise to high tide again.

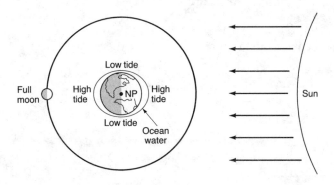

Figure 6-12. High tides occur at positions on Earth facing directly toward or away from the moon. Low tides occur halfway between high tides.

When the sun, Earth, and moon are lined up, the gravity of the sun and moon combine to produce especially high and low tides. When they are not lined up, the tides are less extreme.

EXERCISE 2

---

1. The moon revolves around Earth from one full moon to the next in about one
   (1) month   (2) day   (3) year   (4) week

2. Cheryl wanted to photograph a full moon on January 1, but the night was cloudy. When will be her next opportunity to take a picture of a full moon?
   (1) January 15   (2) January 30   (3) February 10   (4) February 18

3. The moon shines by
   (1) its own light   (3) light reflected off Earth
   (2) radioactivity   (4) reflecting sunlight

4. The diagram shows how the moon looked on three nights in May.

May 1   May 4   May 8

How would the moon most likely appear on the night of May 15?

(1)   (2)   (3)   (4)

5. As the moon revolves around Earth, the moon
   (1) always keeps the same side facing Earth
   (2) does not rotate on its axis
   (3) turns all of its surface toward Earth
   (4) does not reflect sunlight

6. At which position in the moon's orbit can all of the moon's lighted side be seen from Earth?
   (1) *A*
   (2) *B*
   (3) *C*
   (4) *D*

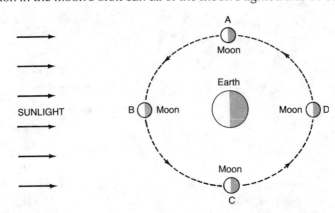

7. A solar eclipse can take place only when the moon is in
   (1) full moon phase
   (2) last quarter phase
   (3) new moon phase
   (4) first quarter phase

8. The diagram below suggests that, in causing tides on Earth, the moon's gravity
   (1) has more effect than the sun's gravity
   (2) has less effect than the sun's gravity
   (3) is equal to the sun's gravity
   (4) has no significant effect

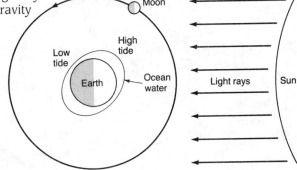

9. The most extreme high and low tides occur when
   (1) the sun, Earth, and moon are at right angles to one another
   (2) there has been a heavy rainstorm
   (3) it is nighttime
   (4) the sun, Earth, and moon are lined up

10. The table at the right shows the times at which high and low tides occurred over two days. Based on the pattern in the table, at about what time will the next low tide occur?
    (1) 12:00 midnight on day two
    (2) 6:00 A.M. on day three
    (3) 12:00 noon on day three
    (4) 6:00 P.M. on day three

Tidal Data Table

Day	Time	Tide
1	4:20 A.M.	Low
	10:35 A.M.	High
	4:45 P.M.	Low
	11:00 P.M.	High
2	5:10 A.M.	Low
	11:25 A.M.	High
	5:35 P.M.	Low
	11:50 P.M.	High

PART III. OUTER SPACE

The Solar System

The **solar system** consists of our sun and all the objects that revolve around it. The major members of the solar system are the sun and the nine planets (see Figure 6-13 on page 94). A number of other objects also come under the influence of the sun's gravity, and so belong to the solar system. These include *satellites* or moons (objects that revolve around planets), asteroids, comets, and meteors.

The Sun

The sun is a hot, bright ball of gases. Nuclear reactions in the sun's interior release enormous amounts of energy, mostly as light and heat. The sun is by far the largest object in the solar system. It is many times larger than Earth. If the sun were hollow, about one million Earths could fit inside.

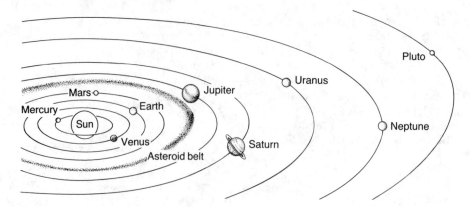

Figure 6-13. The solar system.

The sun is a star, like the stars we see at night. Although the sun is an average-size star, it seems much larger than other stars because it is much closer to Earth. Light from the sun takes about eight minutes to reach Earth. In contrast, light from the nearest star (other than the sun) takes more than four years to reach Earth! The sun is Earth's main source of energy, providing the light and heat necessary for the existence of life.

The Planets

There are nine planets that revolve around the sun. In the night sky, the planets look much like stars. However, as days and weeks go by, planets change position against the background of motionless stars. Also, they do not "twinkle" the way stars do. Unlike stars, planets do not give off their own light. The planets are visible because they reflect the light of the sun.

Each planet has its own special characteristics. Mercury is the closest planet to the sun. Venus is the hottest planet, and the brightest as seen from Earth. Mars is often called the Red Planet, because it reflects reddish light. Jupiter is the largest planet. Saturn is encircled by rings made of rock and ice particles. Pluto is the smallest planet, and usually the farthest from the sun, although part of Pluto's orbit crosses inside Neptune's orbit.

Earth, the third planet from the sun, is unique in that it appears to be the only planet in the solar system that supports life. Much of what we know about the other planets has been discovered with the help of space probes. Table 6-2 on the facing page lists the planets and some facts about them.

Asteroids, Comets, and Meteors

Asteroids are rock fragments of various shapes and sizes that revolve around the sun. They are found mostly in a region called the *asteroid belt*, between the orbits of Mars and Jupiter (see Figure 6-13 above). Scientists think the asteroids are materials left over from the birth of the solar system that never combined to form a planet.

A *comet* is a loose mass of rock, ice, dust, and gases that moves through space as a unit. Comets travel in stretched-out orbits (see Figure 6-14 on the next page). As a comet approaches the sun, energy in sunlight makes the comet glow and produce a "tail." The comet's tail always points away from the sun, regardless of the direction in which the comet is moving.

Table 6-2. Planetary Data

Planet (in order from sun)	Distance from sun in Earth–sun units†	Time to revolve once around sun	Time to rotate once on axis	Number of satellites (moons)
Inner planets:				
Mercury	0.4	88 days	59 days	0
Venus	0.7	225 days	243 days	0
Earth	1.0	365.25 days	24 hours	1
Mars	1.5	1.88 years	24.6 hours	2
Asteroid belt				
Outer planets:				
Jupiter	5.2	11.86 years	9.9 hours	16*
Saturn	9.5	29.63 years	10.6 hours	18*
Uranus	19.2	83.97 years	17 hours	15*
Neptune	30.1	165 years	16 hours	8*
Pluto	39.5	248 years	6.4 days	1

†An Earth–sun unit is the average distance from Earth to the sun (149,600,000 kilometers).
*Number of known moons; the actual number may be higher.

A **meteor** is a fragment of rock traveling through space that enters Earth's atmosphere at high speed. Contact with the atmosphere creates friction that causes the meteor to heat up and burn. This produces a bright streak across the night sky, sometimes called a "shooting star." Occasionally, a large meteor does not burn up completely, and a chunk of rock, called a *meteorite*, reaches Earth's surface.

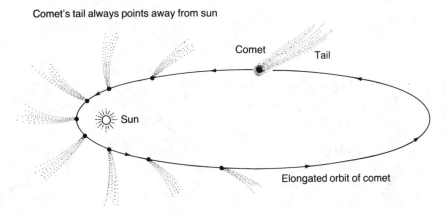

Figure 6-14. Typical orbit of a comet.

The Universe

The *universe* consists of all space and all objects in space. There are far more stars in the universe than there are grains of sand on a beach. Nevertheless, the distances between stars are so vast that most of the universe is empty space.

Stars are not evenly distributed in space, but are clustered together in large groups called *galaxies*. The universe contains billions of galaxies, and each galaxy contains billions of stars. Galaxies are separated by great distances. Our sun is a star in the galaxy called the *Milky Way*. From Earth on a clear night, we can see only a few thousand of the nearby stars in our own galaxy. Astronomers must use telescopes to

PROCESS SKILL: DETERMINING A QUANTITATIVE RELATIONSHIP

For a planet to remain in orbit around the sun, the gravitational pull of the sun on the planet must be balanced by a force pulling the planet away from the sun. Without this balancing force, the planet would not stay in orbit.

The *orbital speed* of a planet is the force pulling it away from the sun. If this speed were increased, the planet would move away from the sun, and if it were decreased, the planet would be pulled in toward the sun. Since the sun's gravity becomes stronger the closer an object gets to it, planets near the sun have to move faster than planets farther away to avoid being pulled into the sun. This suggests that there is a relationship between a planet's orbital speed and its distance from the sun, as shown in the table below.

Planetary Data Table

Planet	Average orbital speed (kilometers per second)	Average distance from sun (kilometers)
Mercury	47.60	57,900,000
Venus	34.82	108,200,000
Earth	29.62	149,600,000
Mars	23.98	227,900,000
Jupiter	12.99	778,000,000
Saturn	9.58	1,427,000,000
Uranus	6.77	2,871,000,000
Neptune	5.41	4,497,000,000
Pluto	4.72	5,913,000,000

The table reveals that as a planet's distance from the sun increases, its orbital speed decreases. This is called an *inverse relationship*, because as one quantity increases (distance from the sun), the other quantity decreases (orbital speed). Answer the questions below.

1. If Earth moved closer to the sun, in order to remain in orbit its orbital speed would have to
   (1) increase
   (2) decrease
   (3) remain the same
   (4) increase first, then decrease

2. If a planet farther from the sun than Pluto were discovered, its orbital speed would be
   (1) faster than Pluto's
   (2) slower than Pluto's
   (3) the same as Pluto's
   (4) impossible to determine

3. The orbital speed of a newly discovered asteroid is found to be 18.61 km/s. The asteroid's orbit is probably located between
   (1) Venus and Earth
   (2) Uranus and Neptune
   (3) Mars and Jupiter
   (4) Saturn and Uranus

see stars in other galaxies. When you look at the night sky, you are seeing only a very small portion of the universe around us.

Distances in Space

Distances in space are so great that they are difficult to fully understand. For example, the distance from Earth to our nearest neighboring star (besides the sun) is about 41,000,000,000,000 kilometers! Because such numbers are so large, astronomers use a unit called a *light-year* to express distances to the stars. A **light-year** is the distance light travels in one year, about 9.46 trillion kilometers. Using light-years allows us to describe distances to stars more conveniently. Our nearest neighboring star, then, is 4.3 light-years away.

EXERCISE 3

---

1. Carol saw a "shooting star" in the night sky. What she actually saw was a
   (1) comet   (2) meteor   (3) planet   (4) moving star

2. Which choice lists solar system objects in order from largest to smallest?
   (1) sun, moon, Jupiter, Earth
   (2) Jupiter, sun, Earth, moon
   (3) Earth, Jupiter, sun, moon
   (4) sun, Jupiter, Earth, moon

3. The diagram shows the four inner planets at various positions in their orbits. Which planet is visible from Earth in the night sky?
   (1) Mercury
   (2) Venus
   (3) Mars

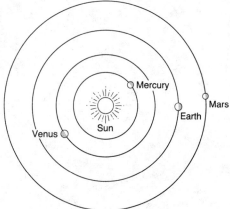

4. The major members of the solar system are
   (1) comets and meteors
   (2) asteroids and satellites
   (3) the sun and the moon
   (4) the sun and the planets

5. The table at the right gives the length of one rotation for five planets. Which planet is spinning most rapidly on its axis?
   (1) Earth
   (2) Venus
   (3) Jupiter
   (4) Mercury

Planet	Length of One Rotation
Mercury	59 days
Venus	243 days
Earth	24 hours
Mars	24.6 hours
Jupiter	9.9 hours

6. Earth is apparently the only planet in our solar system that
   (1) has a moon
   (2) revolves around the sun
   (3) supports life
   (4) has a day side and a night side

7. Examine the column labeled "Distance" in the accompanying star data table. What units of measurement do the numbers in this column most likely represent?
   (1) meters
   (2) kilometers
   (3) miles
   (4) light-years

Star Data Table

Star	Magnitude (Brightness)	Distance
Sirius	1.4	8.8
Vega	0.5	26.4
Capella	−0.6	45.6
Antares	−4.7	423.8

8. Stars in the universe are clustered together in large groups called
   (1) solar systems   (2) galaxies   (3) asteroid belts   (4) constellations

9. Compared to the distances between the planets of our solar system, the distances between stars are
   (1) much greater   (2) much less   (3) about the same   (4) a little greater

10. Jesse saw a group of starlike objects in the night sky on April 7. A month later he noticed that one of the objects had moved to a different position (see the diagram). The object that moved was most likely
    (1) a star
    (2) a planet
    (3) a meteor
    (4) an alien spaceship

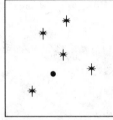

April 7

May 7

Chapter 7. Energy and Motion

PART I. ENERGY BASICS

Force, Work, and Energy

Any push or pull on an object is called a *force*. **Work** is done when a force causes an object to move over a distance. The amount of work done depends on the amount of force applied and the distance the object is moved. The relationship between work, force, and distance is given by the formula:

$$\textbf{work} = \textbf{force} \times \textbf{distance.}$$

When a force is applied to an object, the force may or may not cause the object to move. If the force does not produce motion, no work is done. As shown in Figure 7-1, a force results in work only if motion is produced.

Energy is the ability to do work. For example, gasoline can move a car because gasoline contains energy. All work requires energy.

Kinds of Energy

There are two basic kinds of energy: *potential* and *kinetic*.

(1) *Potential energy* is stored energy that an object has because of its position or its chemical composition. A rock on a cliff top has potential energy because of its position. A lump of coal contains potential energy in its chemical makeup.

Weight is moved over a distance; work is done

Weight held up motionless; no work is done

Figure 7-1. Work is done when a force acts over a distance.

(2) *Kinetic energy* is energy that an object has when it is moving. A rock falling off a cliff has kinetic energy. The heat given off by a burning lump of coal is also a form of kinetic energy. The faster an object moves, the more kinetic energy it has. Figure 7-2 shows examples of potential and kinetic energy.

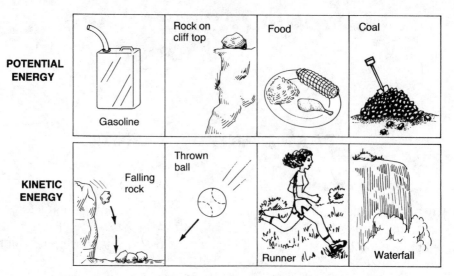

Figure 7-2. Examples of potential energy and kinetic energy.

Potential energy may be changed into kinetic energy when motion is produced. Water held back by a dam has potential energy but no kinetic energy. Releasing the water and letting it flow changes its potential energy into kinetic energy.

Kinetic energy may also be changed into potential energy. When a ball is thrown straight up into the air, its kinetic energy of motion is changed into potential energy as the ball rises higher above the ground. At the ball's highest point, it is motionless and has only potential energy. As the ball falls back to the ground, this potential energy is changed back into kinetic energy.

Forms of Energy

Both potential and kinetic energy exist in many forms. For example, *mechanical energy* is the energy with which moving objects perform work. A hammer striking a nail, a jack lifting a car, and pedals turning the wheel of a bicycle are examples of things using mechanical energy.

Chemical energy is energy stored in certain substances because of their chemical makeup. When these substances are burned, the energy is released. Coal, oil, propane gas, and foods are examples of substances that contain chemical energy.

Magnetic energy is present in certain objects that are rich in iron. Such objects produce a magnetic field that can attract other iron-rich objects. Magnets are often used to help close refrigerator and cabinet doors.

Nuclear energy is the energy stored within the *nucleus* (center) of an atom. This energy can be released by joining atoms together or by splitting atoms apart.

Other important forms of energy include sound, light, heat, and electricity. They are discussed later in this chapter.

Energy Transformations

Energy can be changed, or *transformed*, from one form into another. For instance, when you take a bus to school each morning, chemical energy in gasoline is changed into mechanical energy that turns the wheels of the bus. At school, when the bell rings between classes, electrical energy is transformed into sound energy. And at night, when you turn on a reading light, electrical energy is changed into light energy. Figure 7-3 shows two common energy transformations.

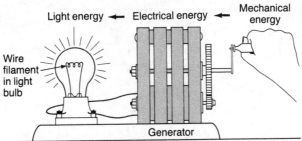

Figure 7-3. The hand-operated generator transforms mechanical energy into electrical energy, which is then transformed into light.

Sometimes, an unwanted form of energy is produced during an energy transformation. For example, a car's motor is designed to change chemical energy into mechanical energy. However, a running motor eventually becomes hot, due to the friction of the motor's parts rubbing against each other. In other words, some of the original chemical energy is transformed into unwanted heat energy.

EXERCISE 1

1. A flowing stream of water is an example of
   (1) potential energy   (2) kinetic energy   (3) nuclear energy   (4) chemical energy

2. For work to be done, a force must produce
   (1) motion
   (2) electricity
   (3) transformation of energy
   (4) potential energy

3. Doing work always requires
   (1) food   (2) chemical reactions   (3) energy   (4) sunlight

4. The best example of an object that possesses potential energy is
   (1) a piece of coal   (2) a falling rock   (3) a rolling ball   (4) a burning log

5. What type of energy is contained in gasoline?
   (1) potential energy   (2) kinetic energy   (3) sound energy   (4) mechanical energy

6. What type of energy change is represented in the diagram below?
   (1) chemical energy to sound energy
   (2) sound energy to chemical energy
   (3) electrical energy to sound energy
   (4) sound energy to electrical energy

7. In the diagram below, a skier is about to start a slide (position *A*), ski down the hill (*B*), and stop at the bottom (*C*). At which position would the skier have the most kinetic energy?
    (1) *A*
    (2) *B*
    (3) *C*

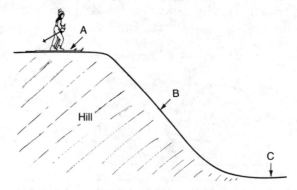

8. The man in the diagram below is pushing on the boulder, but the boulder is not moving. Which statement is true?

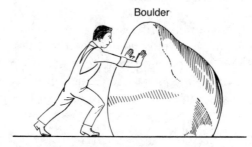

    (1) A force is being applied, and no work is being done.
    (2) A force is being applied, and work is being done.
    (3) No force is being applied, and no work is being done.
    (4) No force is being applied, and work is being done.

9. What type of energy change is shown in the diagram below?
    (1) kinetic energy to potential energy
    (2) nuclear energy to light energy
    (3) chemical energy to heat energy
    (4) chemical energy to mechanical energy

Burning wood

PART II. MACHINES

Machines and Work

A ***machine*** is a device that transfers mechanical energy from one object to another object. Machines make work seem easier. They do this by multiplying force, and by changing the direction or the distance over which a force is applied. For example, a pulley changes the direction of a force. A wrench multiplies the force applied to it when removing a tight bolt. A loading ramp attached to the back of a truck changes the distance over which a force is applied.

The force a machine has to overcome is called *resistance*, and the force applied is called *effort*. Using a machine can reduce the amount of effort needed to overcome a given amount of resistance. However, the amount of work done is not made less by using a machine.

An example will make this clearer. Suppose you had to lift a box weighing 200 pounds (lb.) up onto a platform 5 feet high. To lift the box straight up by yourself, you would need to apply 200 lb. of force over a distance of 5 feet. Using the formula **work = force × distance**, you get:

$$\textbf{work} = 200 \text{ lb.} \times 5 \text{ ft.} = 1000 \text{ ft.-lb.}$$

However, if you set up a rope and a system of pulleys to change the direction and distance of the force required, you might have to pull in 20 feet of rope using only 50 lb. of force:

$$\textbf{work} = 50 \text{ lb.} \times 20 \text{ ft.} = 1000 \text{ ft.-lb.}$$

The pulley system lets you use less effort over a longer distance to do the same amount of work you would have done without the pulleys.

Simple Machines

Many complex modern machines are made up of a number of simple machines working together to perform some task. The *lever* and the *inclined plane* are the most basic simple machines. Most other simple machines are based either on the lever or on the inclined plane.

A *lever* consists of a rigid bar that can turn around a point called a *fulcrum*, as shown in Figure 7-4. Levers make work easier by multiplying applied forces. Examples of levers include a pair of pliers and a crowbar.

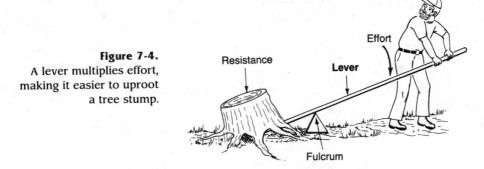

Figure 7-4.
A lever multiplies effort, making it easier to uproot a tree stump.

A *pulley* is a modified form of the lever. Figure 7-5 shows several types of pulleys. A *wheel and axle* is also a modified lever. It consists of a large wheel with a smaller wheel, or axle, in its center. When one wheel is turned, so is the other (see Figure 7-6 on page 104).

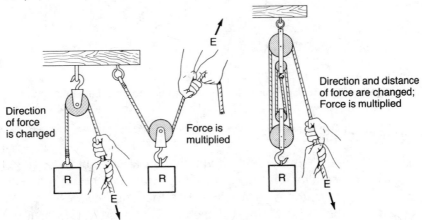

Figure 7-5. Three types of pulleys. (**R** stands for resistance, **E** stands for effort.)

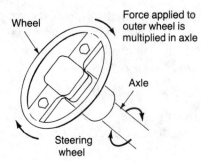

Figure 7-6. A steering wheel is an example of a wheel and axle.

An *inclined plane* is a flat surface with one end higher than the other. A wheelchair ramp is an inclined plane; so is a staircase. Figure 7-7 shows how an inclined plane makes work easier by changing the direction and distance of a force needed to do a job.

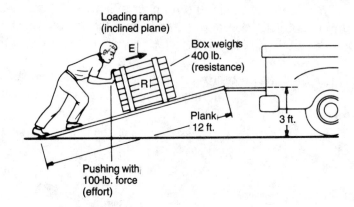

Figure 7-7. A loading ramp is an inclined plane.

The *screw* and the *wedge* are simple machines that are based on the inclined plane (Figure 7-8).

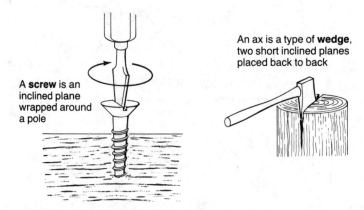

Figure 7-8. The screw and the wedge are based on the inclined plane.

Efficiency of Machines

Ideally, a machine's work output should equal the amount of work put into the machine. However, in reality, machines are never 100 percent efficient—the amount of work done by any machine is always less than the amount of work put into it. This is because some of the work put into a machine is converted into heat energy, and thus wasted. The heat is produced by friction between the machine's moving parts.

Returning to our earlier example, suppose you lift a 200-pound box up 5 feet, using pulleys. Although 1000 foot-pounds of work output was accomplished, you actually had to perform *more* than 1000 foot-pounds of work. Some of your work input is wasted because friction between the wheel and axle of each pulley creates heat.

A machine can be made more efficient by reducing friction. A common way to reduce friction is to *lubricate* the contact surfaces of a machine's moving parts with grease or oil. Other methods include waxing the contact surfaces, sanding the surfaces to make them smoother, or using ball bearings between the surfaces (Figure 7-9).

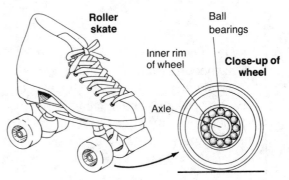

Figure 7-9. Ball bearings in the wheel of a roller skate reduce friction as the wheel turns.

PROCESS SKILL: DESIGNING AN OBSERVATION PROCEDURE

Many common tools are levers of some kind. For example, scissors, shovels, and salad tongs are levers. All levers can be grouped into three basic classes, depending on the location in the lever of the *resistance*, the *effort* applied, and the *fulcrum* (the point around which the lever turns). Diagram 1 illustrates the three lever classes.

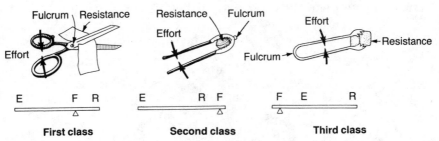

Diagram 1. The three classes of levers.

A *first-class lever*, such as a pair of scissors, has the effort (**E**) applied on one end, the resistance (**R**) on the other end, and the fulcrum (**F**) in between. A *second-class lever*, like a nutcracker, has the fulcrum and effort at opposite ends, and the resistance in the middle. A *third-class lever*, such as a pair of ice tongs, has the resistance and the fulcrum at opposite ends, and the effort applied in the middle.

Diagram 2 on page 106 shows some examples of levers. Can you determine which lever class each item represents? Drawing a *lever diagram* for each one will help you do this. First, draw a line to represent the item. Next, think about how you use the item, and try to iden-

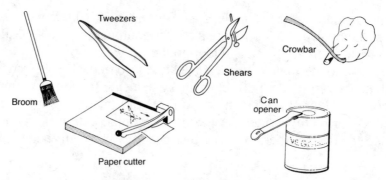

Diagram 2. Examples of levers.

tify the positions of the fulcrum, effort, and resistance. Where does the item meet resistance? Is effort applied to one of the ends of the item, or somewhere in between? Where does the object turn or change direction? Once you have located the fulcrum, resistance, and effort, and labeled them on your lever diagram, you can then classify the item using the definitions of lever classes given above.

Here is an example of how this would be done for a broom:

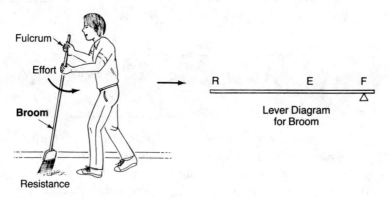

The lever diagram fits the definition of a third-class lever. Do the other items on your own, and then answer the following questions.

1. The wheelbarrow is a type of lever. The wheel in front is the location of the

   (1) effort   (2) fulcrum   (3) resistance

2. Which of the following items is a lever of the type shown in the lever diagram?

   E   R   F

   (1) scissors   (2) paper cutter   (3) tweezers   (4) bottle opener

EXERCISE 2

1. The force that a machine has to overcome is called
   (1) effort   (2) resistance   (3) friction   (4) energy

2. The chain of a bicycle is greased in order to
   (1) increase weight   (2) reduce air drag   (3) reduce friction   (4) increase resistance

3. A screw is a modified
   (1) wheel and axle   (2) lever   (3) pulley   (4) inclined plane

4. A substance that is commonly used to reduce friction between two pieces of metal is
   (1) water   (2) air   (3) oil   (4) sand

5. Andrea moved a heavy box using a pulley system. If her work output was 800 foot-pounds, her work input was
   (1) greater than 800 ft.-lb.
   (2) less than 800 ft.-lb.
   (3) exactly 800 ft.-lb.
   (4) no way to tell

6. The pulley on the flagpole in the illustration makes it easier to raise the flag by
   (1) decreasing the amount of work required
   (2) changing the direction of the force applied
   (3) putting out more work than is put into it
   (4) making the flag lighter

7. Machines transfer which kind of energy from one object to another?
   (1) chemical energy   (2) mechanical energy   (3) potential energy   (4) heat energy

8. What type of simple machine is being used to split the wood?
   (1) wheel and axle
   (2) pulley
   (3) lever
   (4) wedge

9. In science class, Mary tested the efficiency of four machines and recorded the results in a chart. For which machine must she have made an error?
   (1) Lever I
   (2) Lever II
   (3) Lever III
   (4) Pulley

Machine	Efficiency
Lever I	75%
Lever II	100%
Lever III	60%
Pulley	30%

10. What three simple machines are being used in the diagram to help move bricks from position *A* to position *B*?
    (1) inclined plane, wheel and axle, and wedge
    (2) inclined plane, wheel and axle, and lever
    (3) inclined plane, wheel and axle, and screw
    (4) inclined plane, pulley, and lever

PART III. SOUND AND LIGHT

Sound Energy and Sound Waves

Sound is a form of energy produced by a vibrating object. When an object vibrates, it moves rapidly back and forth. This motion pushes and pulls the surrounding air, producing alternating compressed and expanded layers of air particles called *sound waves* (Figure 7-10). These sound waves spread outward in all directions from their source, somewhat like the circular ripples that are produced when you toss a pebble into a calm pool of water.

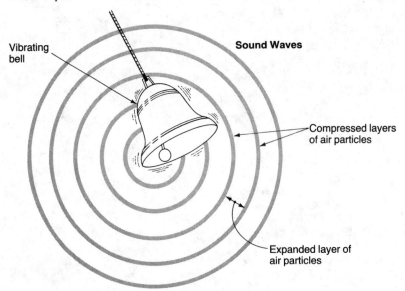

Figure 7-10. A vibrating object produces sound waves.

Objects that can produce sound include bells, radio speakers, guitar strings, or any other thing that can vibrate. For instance, the sound of your voice is caused by vibrating vocal cords in your throat. If you place your hand on your throat while you speak, you can feel the vibrations that produce the sound.

Sound waves can travel only through *matter*, whether in the form of a solid, liquid, or gas. Sound cannot travel through a *vacuum*, an area containing no matter. The substance that sound travels through is called its *medium*.

The Speed of Sound

The speed of sound depends mostly on the density of the substance, or medium, it is passing through. The more dense the medium is, the faster the sound waves can

Table 7-1. Speed of Sound Through
Different Substances (at 25°C)

Medium	State or Phase	Speed (meters/sec)
Iron	Solid	5200
Glass	Solid	4540
Water	Liquid	1497
Air	Gas	346

travel through it. Generally, sound travels fastest through solids, which have the greatest density, and slowest through gases, which have the least. Table 7-1 gives the speed of sound through several substances.

Although the speed of sound can vary, it is always much slower than the speed of light. During a thunderstorm, for instance, a lightning bolt produces a flash of light and a clap of thunder at the same time. The speed of light is so fast that the light reaches us almost instantly. The sound of the thunder travels much more slowly, so we usually hear the thunder after a pause of a few seconds.

Sound Can Be Harmful

Loud noises, especially if prolonged, can damage the hearing abilities of humans and other living things. The damage may be temporary or permanent. Many people work in places where they are exposed to frequent loud noises, such as factories with noisy machinery, construction sites with periodic explosions and loud jackhammers, and even concert halls or recording studios where loud music is played. Workers in such places should protect their hearing by putting earplugs or cotton in their ears.

Light Energy

Light is a visible form of energy. Like sound, light travels in waves that move outward in all directions from their source. But unlike sound, light waves move in straight paths called *rays*, as shown in Figure 7-11. Unlike sound waves, which can curve around corners and objects, light rays cannot curve around objects. This is why objects block out light rays and cast shadows. However, light can travel through a vacuum, something that sound cannot do. In fact, the sun's light travels through the vacuum of space to reach Earth.

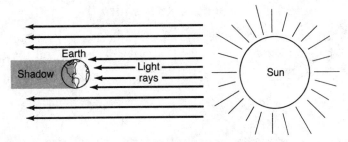

Figure 7-11. Light waves travel in straight paths.

The speed of light is extremely fast, about 300,000 kilometers per second. That is almost a million times faster than the speed of sound! Light that travels over distances we commonly encounter on Earth arrives in just a fraction of a second.

The sun is our main source of light energy. Fire and lightning are other sources of natural light. Light can also be produced artificially, as with a light bulb.

PROCESS SKILL: INTERPRETING GRAPHS

A *line graph* presents information in a visual form that is easy to read. Such graphs usually express a relationship between two changing quantities, or variables. As you know, sound is produced by vibrating objects. The graph below shows the relationship between the speed at which an object vibrates (the number of vibrations per second) and the *pitch* of the sound produced (how high or low it sounds).

A particular pitch, or note, is produced by a specific number of vibrations per second. For instance, the note called "middle C" on the piano is produced by an object vibrating 256 times each second. What happens to the pitch of a sound when the sound vibrations increase in speed? By examining the graph, you can quickly see that as the vibrations speed up, the pitch of the sound rises.

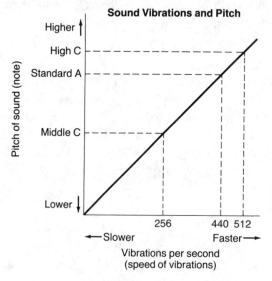

You have learned that sound travels at different speeds through different substances. The speed of sound is faster in solids like stone

Light Can Be Reflected, Absorbed, or Transmitted

When light strikes the surface of an object, three things can happen, as shown in Figure 7-12. Some light may be bounced back, or *reflected*, off the surface. Some

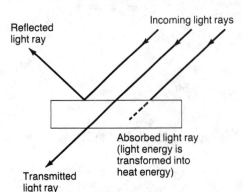

Figure 7-12. When light strikes a surface, some may be reflected, some absorbed, and some transmitted.

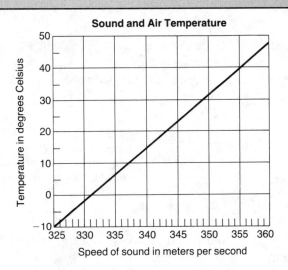

Sound and Air Temperature

Temperature in degrees Celsius (y-axis)

Speed of sound in meters per second (x-axis)

or metal, and slower in liquids and gases, such as water and air. The speed of sound through air is also affected by *air temperature*. The graph above shows the relationship between air temperature and the speed of sound. Study the graph and answer the following questions.

1. Sound travels at a speed of 340 meters per second at about which temperature?
    (1) 22°C (2) 10°C (3) 6°C (4) 15°C

2. At a temperature of 31°C, sound travels at about
    (1) 345 meters per second   (3) 355 meters per second
    (2) 350 meters per second   (4) 340 meters per second

3. What does the graph suggest about the relationship between air temperature and the speed of sound?
    (1) As air temperature decreases, the speed of sound increases.
    (2) As air temperature increases, the speed of sound remains the same.
    (3) As air temperature increases, the speed of sound increases.
    (4) As air temperature increases, the speed of sound decreases.

light may be *absorbed* as heat energy, and some light may be *transmitted*, passing through the object. A shiny metal surface reflects a lot of the light that strikes it. Much of the light striking a blacktopped road is absorbed as heat. Clear glass allows most light to be transmitted through it.

We see objects because their surfaces reflect light. The smoother the surface, the more accurate the reflection. A mirror gives an accurate reflection because it has a smooth, shiny surface. A wall produces a much different kind of reflection because its rougher surface scatters the light.

Objects of different colors absorb light to varying degrees. Dark-colored objects absorb more light as heat energy than do light-colored objects, which reflect more light. For this reason, people usually wear light-colored clothing to keep cool during hot, sunny weather.

Materials also differ in their ability to transmit light. *Transparent* materials, such as window glass, permit almost all of the incoming light to pass directly through them. *Translucent* materials, such as wax paper, let some light pass through, but they scatter the light rays so that images are not transmitted clearly. *Opaque* materials, like wood and iron, do not allow any light to pass through them.

Lenses

Sometimes, when light passes from one transparent substance into another, such as from air into water, the light rays are bent. This is why a pencil in a glass of water looks broken or bent where it enters the water (see Figure 7-13). The light rays being

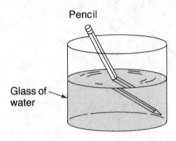

Figure 7-13.
Bending of light rays makes the pencil look bent, or broken.

reflected from the pencil are bent as they pass from the water into the air. This fact has been put to use in the making of lenses.

A *lens* is a piece of transparent glass or plastic that has curved surfaces. The curved surfaces bend light rays that pass through the lens. The shape of a lens determines how it bends light, as shown in Figure 7-14. A lens with surfaces that curve outward bends light rays so that they are focused in toward a common point. A lens with surfaces that curve inward bends light rays so that they spread out.

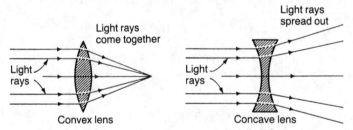

Figure 7-14. The shape of a lens determines how it bends light.

Images of objects seen through lenses may be larger, the same size, or smaller than the object itself. For instance, the lens of a camera forms smaller images of objects. A photocopy machine has a lens that forms images the same size as the original object. Binoculars contain lenses that magnify objects, making them appear larger.

The Electromagnetic Spectrum

Light waves are part of a larger family of energy waves that can travel through a vacuum at the speed of light. These are called *electromagnetic waves*. They include radio waves, microwaves, infrared waves, visible light, ultraviolet rays, X rays, and gamma rays. Together, these energy waves make up the *electromagnetic spectrum* (see Figure 7-15 on the facing page).

We are all familiar with some uses of electromagnetic waves. Radio waves are used for radio and television broadcasting. Microwaves are used in communications and in microwave ovens. X rays are used to diagnose illnesses and injuries.

Overexposure to electromagnetic waves can be harmful to living things. We should be especially careful to limit our exposure to certain types of electromagnetic radiation, such as X rays and ultraviolet rays, which have been linked to cancer.

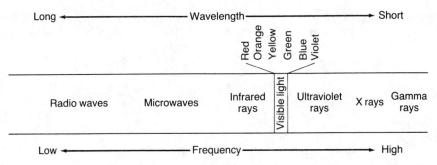

Figure 7-15. The electromagnetic spectrum.

EXERCISE 3

1. Compared to the speed of light, the speed of sound is
   (1) much faster   (2) much slower   (3) the same   (4) a little slower

2. The bar graph below shows the average speed of sound through solids, liquids, gases, and a vacuum. Based on the graph, through which of the following would sound travel fastest?
   (1) air
   (2) rock
   (3) water
   (4) outer space

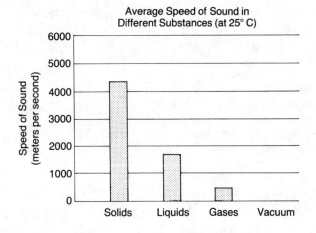

3. Sound waves can travel through
   (1) water only   (2) air only   (3) all matter   (4) a vacuum

4. A blast of dynamite set off by a work crew produced a bright flash of light and a loud explosion. A person standing 2 kilometers away with a clear view of the work site would
   (1) hear the sound first, then see the flash
   (2) see the flash first, then hear the sound
   (3) see the flash and hear the sound at the same time
   (4) hear the sound, but not see the flash

5. Lenses can produce images that are
   (1) larger than the original object
   (2) the same size as the original object
   (3) smaller than the original object
   (4) all of the above

6. We can see objects mainly because they
   (1) bend light   (2) reflect light   (3) absorb light   (4) transmit light

7. When using the apparatus shown in the diagram below, the student could see the flame only if all three holes were lined up. What property of light does this demonstrate?

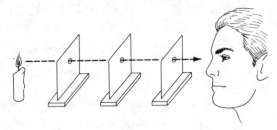

   (1) Light rays are reflected from smooth, shiny surfaces.
   (2) Light rays are absorbed as heat by dark-colored surfaces.
   (3) Light rays travel in straight paths.
   (4) Transparent objects transmit most of the light that strikes them.

8. The accompanying diagram shows three ways that light can behave when striking a sheet of colored glass. The light rays at location *C* have been
   (1) reflected by the glass
   (2) transmitted by the glass
   (3) absorbed by the glass
   (4) blocked out by the glass

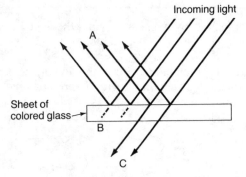

9. X rays are used for
   (1) radio and television broadcasting
   (2) communications and cooking food
   (3) diagnosing illnesses and injuries
   (4) all of the above

10. Overexposure to electromagnetic waves
   (1) has no effect on living things
   (2) always has a good effect on living things
   (3) can have harmful effects on living things
   (4) cannot occur in a vacuum

PART IV. HEAT AND ELECTRICITY

Heat Energy

All matter is composed of particles that vibrate in constant motion. *Heat energy* is the energy of motion of these particles. When the particles of a substance are vibrating slowly, the substance is relatively cold. When the particles of a substance are in rapid motion, the substance is hot.

When a material is heated, it expands. When heat energy is removed from a substance, the opposite takes place: the material contracts, or shrinks. Both expansion and contraction involve an actual change in the size (volume) of the material.

For instance, bridges and railroad tracks contain thin spaces that allow the solid materials they are made of to expand in warm weather without damaging the overall structure. Liquids and gases also respond to heating and cooling in this way. Thermometers contain a liquid that expands and contracts inside a tube, providing a measurement of heat. A volleyball filled with enough air to feel firm to the touch on a warm summer day will feel "flabby" on a cold winter day, because the air inside the ball contracts in response to the cold.

Almost all substances expand when heated and contract when cooled. Water is an exception. At most temperatures, water behaves like other substances. However, when water is cooled to near its freezing point, it stops contracting and begins expanding. In fact, ice takes up more space than the liquid water that freezes to form the ice. This is why an unopened bottle of water will crack if left outdoors during freezing weather. The force generated by the expansion of water changing to ice is so powerful that it can crack glass, rocks, concrete, and even metal.

Heat Transfer

Heat always flows from warmer objects or places to cooler objects or places. Some materials transfer, or *conduct*, heat from warmer to cooler areas quickly and easily. For instance, a metal spoon put into a cup of hot tea quickly gets hot because metal conducts heat well. Figure 7-16 shows how heat is conducted along a metal rod.

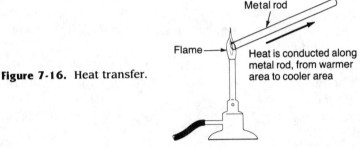

Figure 7-16. Heat transfer.

On the other hand, materials that do not transfer heat well can be used as *insulation* to reduce the flow of heat from warm areas to cooler areas. For example, in winter, a curtain in front of a window acts as insulation to reduce the flow of heat from the warm air inside a room to the cold air outside. Also, a pot holder reduces heat flow so you can grasp the metal handle of a hot frying pan without being burned.

Electrical Energy

Electricity is a form of energy produced by the flow of electrons from one point to another. *Electrons* are tiny particles that orbit around the nucleus of an atom. Humans have found many uses for electrical energy. For instance, the electricity in your home can be used to power light bulbs, air conditioners, television sets, and many other appliances.

Electrical **conductors** are materials through which electricity moves easily. Conductors provide a ready path for the flow of electrons. Most metals are good conductors. Substances that are poor conductors and do not allow electrons to flow through them are called *insulators*. Rubber, glass, and plastic are common insulators.

Electrical wires in your home consist of a conductor, like copper or aluminum, wrapped in a protective coating of an insulator, like rubber or plastic (see Figure 7-17). These wires provide a safe path for electricity.

Figure 7-17. An electrical wire consists of a conductor (metal) wrapped in an insulator (plastic).

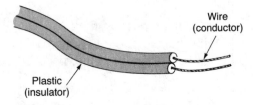

PROCESS SKILL: USING LABORATORY EQUIPMENT SAFELY

When doing laboratory work that requires heat, a number of safety precautions should be followed:

1. Safety goggles should always be worn when heating liquids, because they can spatter in your face and eyes.
2. Insulated gloves should always be worn when you are moving hot objects, unless the object has insulated handles.
3. You should always be aware of people around you when you move anything hot.
4. Sources of heat can sometimes cause problems:
   a. Liquid fuels can spill and catch on fire; they should be handled carefully.
   b. Hot plates may not always look hot. Never leave a hot plate unattended while it is on, and always turn it off immediately when you are finished using it.
   c. Gaseous fuels can cause explosions. Always wear safety goggles when using a gas jet, and be sure to turn gas jets off when you are through using them.
5. A fire extinguisher should be nearby whenever you are working with flames.

What safety precautions should be taken when doing the activity shown in Diagram 1 below? When preparing hydrochloric acid, safety goggles and insulated gloves should be worn. The person performing the activity should be aware of other people in the area. Finally, there should be a fire extinguisher nearby.

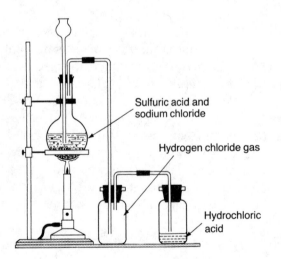

Sulfuric acid and sodium chloride

Hydrogen chloride gas

Hydrochloric acid

Diagram 1.
Laboratory preparation of hydrochloric acid.

Diagram 2 on the facing page shows a setup used to determine what happens to a liquid's volume when it is heated. This experiment has been poorly set up, because too rapid heating of the liquid may

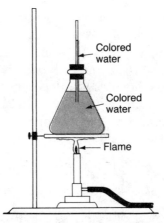

Diagram 2.
Effect of heat on
a liquid's volume.

cause it to spurt out the top of the tubing. Examine the setup and use your common sense to answer the following questions.

1. A safer way to heat the flask would be to
    (1) use a shorter piece of glass tubing
    (2) use more gas to make the flame bigger and hotter
    (3) use a greater amount of water in the flask
    (4) suspend the flask in a beaker of water so that the flame is not directly on the flask

2. A safe way to quickly remove the flask from the heat source would be to
    (1) remove it with your bare hands
    (2) remove it while wearing insulated gloves
    (3) remove it with metal tongs
    (4) remove it by the glass tubing

Electric Circuits

An *electric circuit* is a complete path for the flow of electricity. A circuit must contain a source of electrical energy, a conducting path for the flow of electrons, and an electrical device that uses the current. The source of electricity could be a battery or a generator. Wires usually provide the path for the electricity in a circuit, and the device could be a lamp or a stereo.

At least two wires are needed for a complete circuit, one wire to carry a flow of electrons to the device and another to carry electrons back to the source of electricity. In addition, a circuit often includes a switch that turns the flow of electricity on and off. Figure 7-18 shows a simple electric circuit with an on–off switch.

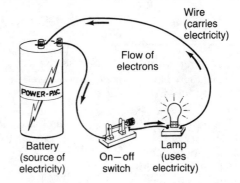

Figure 7-18.
A simple electric circuit, with an
on–off switch included.

Safe Use of Electricity

Electricity is dangerous and can injure or kill living things. Though it has become a common part of our lives, electricity should always be handled carefully.

When too many appliances are using electricity from one circuit at the same time, the circuit can become *overloaded*. This makes the wires heat up, which can cause an electrical fire. **Fuses** and **circuit breakers** prevent overloading of circuits by automatically interrupting the flow of electricity when it reaches a dangerous level. These protective devices are essential in a home wiring system. Figure 7-19 shows how a fuse works.

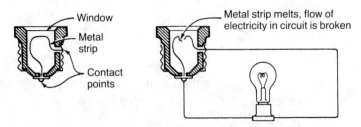

Figure 7-19. How a fuse works: When the electric circuit begins to overheat, the metal strip in the fuse melts, breaking the flow of electricity.

Grounding of electrical appliances is another important safety precaution. Sometimes an electrical charge can build up in an electrical device. A person coming in contact with the device may receive a severe shock from this excess charge. Grounding helps prevent such accidents.

To ground an electrical device, a wire is run from the device to a conducting material, such as a metal pipe, that is in contact with the earth. This provides a safe outlet for any excess charge in the device by conducting it to the earth, where it is absorbed. Some appliances come equipped with a three-pronged plug (Figure 7-20). The third prong provides grounding and should always be used.

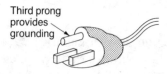

Third prong provides grounding

Figure 7-20. A three-pronged plug.

You should observe common-sense safety rules whenever using electricity. Always unplug electrical devices before cleaning or repairing them. Electrical wires should not be used if they are broken or frayed, and plugs should be replaced if the wire is broken near the plug. Never handle electrical appliances while you are in water or if you are wet, because water can act as a conductor.

EXERCISE 4

---

1. Water differs from most other substances in that it
   (1) expands when heated
   (2) contracts when cooled
   (3) expands when it freezes
   (4) contracts when it freezes

2. A fire burning in a fireplace warms the air in a cold room because
   (1) the fire causes air particles to move more slowly
   (2) the room is not well insulated
   (3) heat always flows from warmer areas to cooler areas
   (4) warm air sinks and cool air rises

3. Placing a metal spoon with a temperature of 20°C into a cup of water with a temperature of 90°C will cause the spoon to
   (1) increase in temperature
   (2) decrease in temperature
   (3) remain the same temperature
   (4) contract in size

4. A thermos bottle keeps warm liquids warm and cold liquids cold. A thermos bottle must be made from a good
   (1) heat conducting material
   (2) heat insulating material
   (3) heat expanding material
   (4) heat contracting material

5. A thermometer works on the principle that, when heated, a substance will
   (1) contract   (2) expand   (3) remain the same size   (4) give off light

6. Which graph best shows the relationship between the temperature of a substance and the motion of the particles in the substance when it is heated?

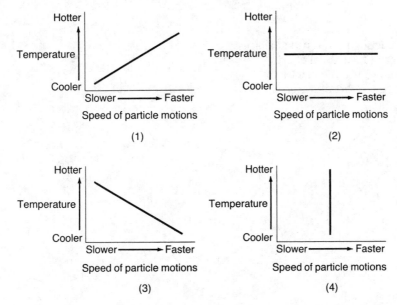

7. The end of a metal bar is placed in a flame for five minutes. The temperature is measured by thermometers at four points on the bar, as shown in the diagram below. The lowest temperature will most likely be recorded at thermometer
   (1) *A*
   (2) *B*
   (3) *C*
   (4) *D*

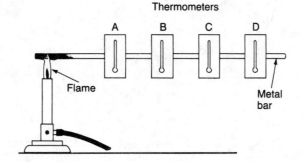

8. A good conductor of electricity is
   (1) plastic   (2) rubber   (3) aluminum   (4) glass

9. Grounding is important because it
   (1) stops and starts the flow of electricity in a circuit
   (2) helps prevent fires caused by overloaded circuits
   (3) insulates metal wires
   (4) helps prevent shocks by conducting excess electricity to the earth

10. The simple electric circuit shown below is incomplete. What item is missing?

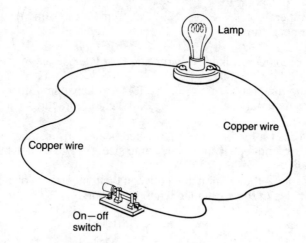

(1) a conducting path for the electricity
(2) a source of electrical energy
(3) a device to use the electricity
(4) a switch to start and stop the flow of electricity

Chapter 8. The Chemistry of Matter

PART I. SAFETY WITH CHEMICALS

The Chemicals Around Us

All the materials of the natural world are made of chemicals. We humans are composed of chemicals, as are all the things around us, including rocks, water, air, food, plants, and animals.

Dangerous Chemicals

Most of the chemicals we encounter are harmless. However, some chemicals used at home and on the job can be dangerous, and must be handled carefully. Household substances that can be hazardous include bleach, ammonia, lye, paint thinner, some kinds of glue, and all medications.

Among the most dangerous household chemicals are those that give off *vapors* (fumes or gases). These vapors may be flammable, toxic, or both. A *flammable* substance is one that burns readily. Flammable vapors may explode from the touch of a spark. Gasoline and most paint thinners are highly flammable and give off flammable vapors.

Toxic substances are poisonous. Breathing toxic vapors can cause death or serious illness. Ammonia, paint thinners, and some kinds of glue produce toxic vapors. Chemicals that give off toxic or flammable vapors should be used only in well-ventilated areas (areas with a constant flow of fresh air) and away from sparks or flames.

Products containing dangerous chemicals are required by law to carry a warning label. A typical label might read, **"DANGER! Contents flammable. Do not use near flame."** If the product is toxic, it may carry the "skull and crossbones" symbol, shown in Figure 8-1.

Figure 8-1.
The "skull & crossbones" symbol on a warning label indicates a toxic substance.

Some substances are dangerous to touch. A *caustic* substance can destroy living tissue, such as skin. A *corrosive* substance will eat through skin, clothing, and most metals. Many oven cleaners and drain openers contain lye, an extremely caustic substance. Acid from car batteries is very corrosive. Keep these chemicals away from your skin and especially your eyes.

Some household chemicals become particularly dangerous when mixed with other chemicals. For instance, although bleach and ammonia are both cleansers, when combined, they produce an extremely toxic gas. Never mix chemicals together unless you are instructed to do so, since combining them may produce a hazardous substance, or even an explosion.

Pressurized Containers

Many household substances are stored under high pressure in spray cans. Deodorants, hair sprays, spray paints, and even whipped cream are sold in pressurized containers. These cans may explode if punctured or heated. A typical label on a spray can might read, **"Warning! Contents under pressure. Do not puncture or incinerate [burn] can. Do not store at temperatures above 120°F."** Such warnings should always be taken seriously.

Playing It Safe

Even "safe" chemicals are safe only if used properly. For example, aspirin is usually a safe, beneficial drug, yet children have died from aspirin overdoses. **All drugs and chemicals should be stored out of the reach of children and pets, preferably in a locked cabinet.**

Before taking any medication or using any chemical product, always read the label carefully and follow the instructions. Remember that *all* chemicals can be dangerous if used improperly. Table 8-1 lists some common household chemicals and their dangers.

Table 8-1. Household Substances and Their Dangers

Substance	Dangers	Special Precautions
Ammonia	Toxic; gives off toxic vapors	Do not mix with other chemicals; use in well-ventilated area.
Chlorine bleach	Toxic; gives off toxic vapors	Do not mix with other chemicals; use in well-ventilated area.
Gasoline	Toxic and flammable; gives off toxic and highly explosive vapors	Keep away from flame; carry only in special containers.
Hair spray	Flammable vapors; pressurized can	Do not puncture or incinerate can; do not store at high temperature.
Lye (in oven cleaners and drain openers)	Toxic and caustic	Wear rubber gloves; protect eyes; do not mix with other chemicals.

Safety in the Laboratory

Special precautions must be observed in the laboratory when handling chemicals and laboratory equipment. There is always a danger of explosions, breakage, spills, and spattering. Any chemical spilled on your skin, clothing, or desk should immediately be washed away with plenty of water.

Even water becomes dangerous when heated. For this reason, safety goggles must be worn in the laboratory at all times. To protect your body and clothing, a rubber apron can also be worn. Long hair should be tied back. When pouring liquids or transferring solids, be careful to avoid spills and to keep the chemicals pure.

When testing any unknown substance, always assume that it may be dangerous. For example, to observe the odor of an unknown substance, fan the vapors toward

Never inhale
vapors directly

Fan vapors toward
nose and sniff gently

Wrong!

Right!

Figure 8-2. Testing the odor of an unknown substance.

your nose and sniff *gently*. Do *not* put your nose directly in the vapors or inhale deeply (see Figure 8-2). Also, *never* taste a chemical.

EXERCISE 1

1. Materials of the natural world that are made up of chemicals include
   (1) living things only
   (2) nonliving things only
   (3) both living and nonliving things

2. Which liquid produces flammable vapors?
   (1) water (2) bleach (3) ammonia (4) gasoline

3. Which substance produces toxic vapors?
   (1) ammonia (2) aspirin (3) water (4) perfume

4. A substance that is dangerous to touch is best described as
   (1) flammable (2) caustic (3) poisonous (4) toxic

5. If a warning label says "use only in a well-ventilated area," this caution probably indicates that the product
   (1) is corrosive
   (2) is under pressure
   (3) gives off dangerous vapors
   (4) has no dangers

6. The warning "Do not puncture or incinerate" is most likely to be found on
   (1) a can of hair spray
   (2) a bottle of ammonia
   (3) a container of milk
   (4) a bottle of aspirin

7. All gasoline stations in New York State have "No Smoking" signs posted. This is because
   (1) smoking is bad for your health
   (2) gasoline is highly toxic
   (3) gasoline is highly corrosive
   (4) gasoline is highly flammable

8. You should wear rubber gloves when using oven cleaners because they contain chemicals that are
   (1) highly toxic (2) caustic (3) under pressure (4) flammable

9. Which chemical is most damaging to the skin?
   (1) alcohol   (2) household ammonia   (3) lye   (4) gasoline

10. When you heat a test tube of water in the science laboratory, what is the most important safety precaution?
    (1) Use cold water.   (3) Wear safety goggles.
    (2) Tilt the test tube.   (4) Use a low flame.

PART II. MATTER

Defining Matter

Look around you. The objects you see, such as this book, your desk and chair, and the walls and ceiling, are all composed of *matter*. The air that surrounds you, which is a mixture of gases, is also made of matter. In fact, every solid, liquid, and gas is a form of matter.

Matter is defined as anything that has *mass* and takes up space. *Mass* is the total amount of material in an object. We measure mass with an equal-arm balance, as shown in Figure 8-3. Notice that a balloon filled with air has a greater mass than an empty balloon, since air has mass. The amount of space an object occupies is called its *volume*. The air in the filled balloon in Figure 8-3 takes up space, giving the balloon a greater volume than the empty balloon.

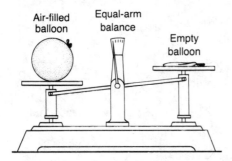

Figure 8-3. The air-filled balloon is heavier and takes up more space than the empty balloon because air is matter.

Is there anything that is not made of matter, that has no mass and takes up no space? Figure 8-4 shows that shining a light on a balance has no effect on the balance. This is because light is a form of energy. Energy is not matter, since it has no mass and no volume. Some other forms of energy are heat and sound.

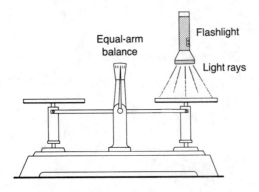

Figure 8-4. The balance is unaffected by the light shining on it, because light is not matter.

PROCESS SKILL: MAKING MEASUREMENTS; PERFORMING CALCULATIONS

The volume of a liquid is often measured with a *graduated cylinder*. This device can also be used to determine the volume of small solid objects, as shown in the diagram below. When an object such as a glass marble or a small rock is placed into the water inside the cylinder, the water level rises. The change in the volume of the water is equal to the volume of the object.

For example, in the diagram, the graduated cylinder contains 20 milliliters (mL) of water. When the marble is placed in the cylinder, the water level rises to 25 mL. The change in the water's volume is 5 mL, so the volume of the marble must be 5 mL. Study this procedure, and answer the following questions.

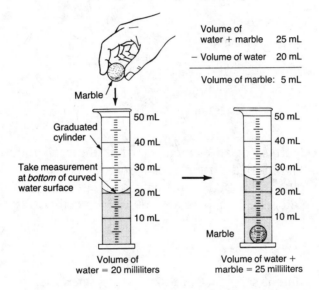

1. The diagram below indicates that the volume of the rock is
   (1) 30 mL
   (2) 20 mL
   (3) 10 mL
   (4) 5 mL

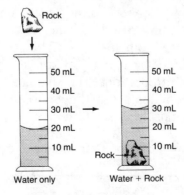

2. A graduated cylinder contains 20 mL of water. If a stone with a volume of 12 mL and another stone with a volume of 5 mL are both placed into this cylinder, the water level will rise to
   (1) 32 mL (2) 25 mL (3) 17 mL (4) 37 mL

Elements

The basic building blocks of matter are called **elements**. All substances are made up of one or more elements. Oxygen, hydrogen, gold, and iron are examples of elements. Each element is represented by a symbol made up of one or two letters. For example, the symbol for hydrogen is **H**, oxygen is **O**, and gold is **Au**. There are 109 known elements. However, fewer than half of them occur commonly in nature. Table 8-2 lists the most common elements found in Earth's crust.

Table 8-2. Most Common Elements in Earth's Crust

Element	Chemical Symbol
Oxygen	O
Silicon	Si
Aluminum	Al
Iron	Fe
Calcium	Ca
Sodium	Na
Potassium	K
Magnesium	Mg

The smallest particle of an element that has the properties of that element is called an **atom**. All atoms of a particular element are alike, but they are different from the atoms of any other element. For instance, all hydrogen atoms are alike, but they differ from oxygen atoms. Since there are 109 elements, there are 109 different kinds of atoms.

Compounds

Millions of different substances are known to scientists. How is this possible if there are only 109 elements? Elements can combine to form new substances. A substance formed when two or more different elements combine is called a *compound*. Since many different combinations of elements are possible, many different compounds can exist. The common substance water is a compound formed when the elements hydrogen and oxygen combine.

A compound is represented by a chemical formula that indicates which elements have combined, and in what proportions. The chemical formula for water, H_2O, indicates that water contains two atoms of hydrogen to every atom of oxygen. Table 8-3 lists some common compounds and their formulas.

Table 8-3. Some Common Compounds and Their Chemical Formulas

Compound	Formula	Elements
Table salt	NaCl	Sodium, Chlorine
Water	H_2O	Hydrogen, Oxygen
Sugar (sucrose)	$C_{12}H_{22}O_{11}$	Carbon, Hydrogen, Oxygen
Quartz	SiO_2	Silicon, Oxygen
Ammonia	NH_3	Nitrogen, Hydrogen

The smallest particle of a compound is called a *molecule*. A water molecule is composed of two hydrogen atoms and one oxygen atom, as shown in Figure 8-5. Atoms of the same element can also combine to form molecules. For example, two oxygen atoms combine to form a molecule of oxygen gas, O_2.

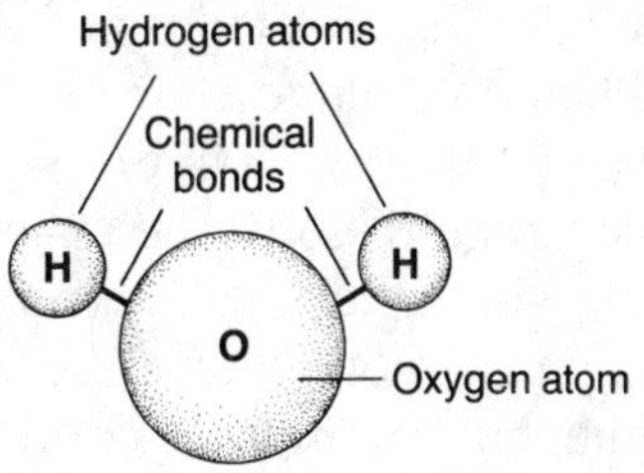

Figure 8-5.
The arrangement of atoms in
a molecule of water.

Atoms and molecules are extremely small. To get an idea of just how small, consider that one teaspoonful of water contains about 175 *sextillion* water molecules. (That would be written as 175 followed by 21 zeroes!)

Chemical Bonds

Atoms in a molecule are joined together by a special link called a **chemical bond** (see Figure 8-5 above). These bonds contain chemical energy. Sometimes, this energy can be released by a *chemical reaction*.

Burning is one type of chemical reaction that releases energy. When wood is burned, energy stored in the chemical bonds within the wood is released as heat and light. Respiration is another chemical reaction that releases energy from bonds.

Physical Properties

Have you ever mistaken salt for sugar? To the eye, they look very much alike. How might you tell them apart? Scientists faced with similar problems identify substances by examining their *properties.*

A difference in taste would help you distinguish salt from sugar. A difference in color helps distinguish salt from pepper. Taste and color are **physical properties**—properties that can be determined without changing the identity of a substance. All substances have unique physical properties by which they can be identified. Table 8-4 lists some physical properties often used for this purpose.

Table 8-4. Examples of Physical Properties

Property	Example
Phase	Mercury is a liquid at room temperature.
Color	Sulfur is yellow.
Odor	Hydrogen sulfide smells like rotten eggs.
Density	Lead is much denser than aluminum.
Solubility	Salt dissolves in water.
Melting point	Ice melts at 0°C.
Boiling point	Water boils at 100°C.

Phases

One obvious physical property of a substance is whether it is a solid, a liquid, or a gas. These three forms of matter are called **phases**. The phase of a substance is determined by the arrangement and motion of the molecules within it.

(1) In *solids*, the molecules are close together, move relatively slowly, and remain in *fixed* (unchanging) positions. A solid has a definite shape and a definite volume; that is, its shape and size do not depend on the container it is in.

(2) In *liquids*, the molecules are usually farther apart and faster moving than the molecules in solids. The molecules in a liquid can change position and flow past each other. A liquid has no definite shape; it takes on the shape of its container. However, liquids do have a definite volume. A given quantity of a liquid takes up the same amount of space, regardless of the shape of its container.

(3) In *gases*, the molecules are much farther apart and move even faster than in liquids, and they can move anywhere within their container. A gas has no definite shape or volume, but expands or contracts to fill whatever container it is in. Figure 8-6 shows how molecules are typically arranged in solids, liquids, and gases.

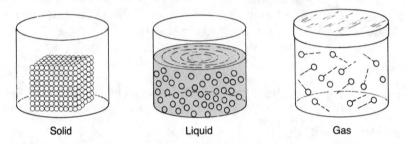

Solid   Liquid   Gas

Figure 8-6. The three phases of matter: solid, liquid, and gas.

Changes in Phase

Since the phase of a substance depends on the arrangement of its molecules, a change in this arrangement can bring about a change in phase.

(1) To change a solid into a liquid, the molecules must generally be moved farther apart, out of their fixed positions. This is called **melting**. Energy must be added to a substance to separate its molecules, so energy is absorbed during melting.

(2) **Freezing** is the opposite of melting. When a liquid freezes into a solid, the molecules come together and bond more tightly into fixed positions. This process releases energy.

(3) Changing a liquid into a gas, by **boiling** or **evaporation**, requires that the molecules of the liquid be separated even further. Energy is therefore absorbed when a liquid changes into a gas.

(4) The change from a gas to a liquid is called **condensation**. During condensation, molecules of a gas move closer together to form a liquid, and energy is released. Figure 8-7 illustrates the energy changes associated with changes in phase.

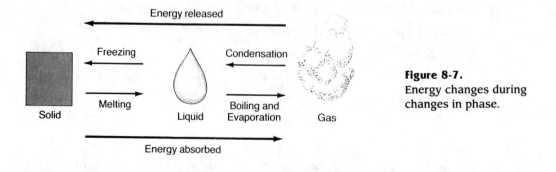

Figure 8-7.
Energy changes during
changes in phase.

For each substance, the change in phase from solid to liquid occurs at a particular temperature called its *melting point*. The melting point of ice, the solid form of water, is 0°C, or 32°F. The temperature at which a liquid freezes into a solid is called its *freezing point*. The freezing point of water is 0°C, or 32°F. The freezing point and melting point of a substance are always the same.

The temperature at which a liquid boils and changes rapidly into a gas is called its *boiling point*. The boiling point of water is 100°C, or 212°F. This is also the temperature at which water vapor begins to condense into liquid water.

PROCESS SKILL: INTERPRETING DATA

By using melting point and boiling point information, we can determine what phase a substance will be in at a given temperature. If the temperature of a substance is below its melting point, the substance is a solid. If its temperature is above its boiling point, the substance is a gas. If the temperature is between the melting and boiling points, the substance is a liquid. For example, at room temperature (20°C), water is a liquid, because 20°C is between the melting point and boiling point of water.

The table below lists the melting points and boiling points of some common substances. What phase would table salt be in at a temperature of 1000°C? Since 1000°C is above the melting point of salt, but below its boiling point, table salt would be a liquid at that temperature. Use the same kind of reasoning to answer the following questions.

Melting Points and Boiling Points of Some Common Substances

Substance	Melting Point (°C)	Boiling Point (°C)	Phase at Room Temperature (20°C)
Water	0	100	Liquid
Alcohol	−117	78	Liquid
Table salt	801	1413	Solid
Oxygen	−218	−183	Gas

1. At a temperature of −190°C, oxygen is in the form of a
   (1) gas (2) liquid (3) solid

2. If alcohol is in the form of a liquid, which of the following could *not* be its temperature?
   (1) −100°C (2) 32°C (3) 100°C (4) 77°C

3. The only substance listed that could be a liquid at a temperature of 80°C is
   (1) table salt (2) water (3) alcohol (4) oxygen

EXERCISE 2

1. What does the diagram below show about matter?
   (1) Matter is made up of elements.
   (2) Matter takes up space.
   (3) Matter is a solid.
   (4) Matter has mass.

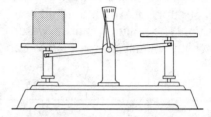

2. Which is not an example of matter?
   (1) water   (2) air   (3) gold   (4) sound

3. The amount of space an object occupies is called its
   (1) volume   (2) mass   (3) weight   (4) length

4. Atoms in a molecule are joined together by
   (1) chemical bonds   (3) electricity
   (2) magnetism   (4) gravity

5. The circles in the closed jars shown below represent particles of matter. Which jar most likely contains a solid?

      (1)   (2)   (3)   (4)

6. When you pour water from a beaker into a flask, there is a change in its
   (1) mass   (2) volume   (3) shape   (4) density

7. In which phase of matter are the particles farthest apart and moving the fastest?
   (1) solid   (2) liquid   (3) gas

8. Which change in phase releases energy?
   (1) solid to liquid   (3) liquid to solid
   (2) liquid to gas   (4) solid to gas

9. Condensation refers to a phase change from
   (1) solid to liquid   (3) liquid to gas
   (2) liquid to solid   (4) gas to liquid

10. The temperature at which a substance melts is the same as the temperature at which it
    (1) boils   (2) freezes   (3) condenses   (4) evaporates

PART III. CHANGES IN MATTER

Physical Changes

As you know, the chemical formula for water is H_2O. What is the formula for ice? When water freezes, the arrangement of its molecules changes, but the molecules themselves do not change. They are still H_2O. A change of phase, such as freezing or melting, does not produce any new substances. A change that does not result in the formation of any new substances is a ***physical change***. All changes of phase are physical changes. Crushing ice cubes into small pieces is also a physical change, since both crushed ice and ice cubes are made of the same substance.

Similarly, when you dissolve sugar in water, the sugar still tastes sweet and the water is still wet. No new substances have been formed, so dissolving is a physical change. Figure 8-8 shows why boiling, melting, and dissolving are physical changes.

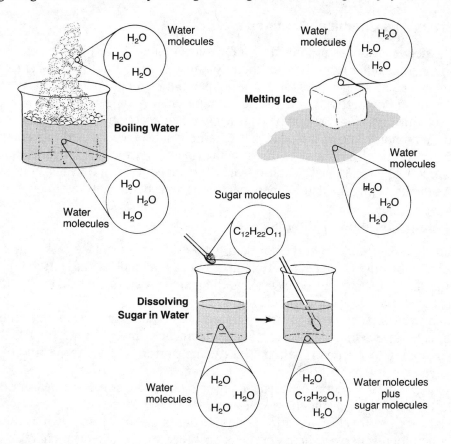

Figure 8-8. During physical changes, no new substances are formed.

Chemical Changes

What happens if you forget to put a carton of milk back into the refrigerator? First, the milk gets warm. This is a physical change. However, if you leave the milk out too long, it turns sour. The sour taste is caused by the production of a new substance called *lactic acid*. A change that produces one or more new substances is called a ***chemical change***, or a chemical reaction. Burning paper produces smoke and ash. Burning is always a chemical change. Table 8-5 on page 132 gives some examples of chemical changes.

Table 8-5. Examples of Chemical Changes

Chemical Change	Original Substance(s)	Substance(s) Formed
Burning coal	Carbon (C) Oxygen gas (O_2)	Carbon dioxide gas (CO_2)
Rusting of iron	Iron (Fe) Oxygen gas (O_2)	Rust (Fe_2O_3)
Tarnishing of silver	Silver (Ag) Sulfur (S)	Tarnish (Ag_2S)
Electrolysis of water	Water (H_2O)	Hydrogen gas (H_2) Oxygen gas (O_2)
Photosynthesis	Carbon dioxide gas (CO_2) Water (H_2O)	Glucose ($C_6H_{12}O_6$) Oxygen gas (O_2)

Properties and Chemical Changes

The new substances produced by a chemical change have their own properties. These properties differ from those of the original substances that reacted, since those substances are no longer present. For example, the element sodium is a metal that explodes upon contact with water. The element chlorine is a poisonous green gas. When sodium and chlorine combine in a chemical reaction, they produce sodium chloride, commonly known as table salt. The new substance formed has completely different properties from those of the original materials, which no longer exist.

During a chemical reaction, atoms are rearranged to form new substances. This involves the breaking of existing chemical bonds and the formation of new bonds.

Energy and Chemical Changes

Simply mixing sodium and chlorine together does not produce table salt. Energy is needed to start the chemical reaction. Likewise, a match does not start to burn until you strike it. The friction caused by striking the match provides the heat energy needed to start the chemical reaction of burning.

Many chemical changes must be started by the addition of energy, in the form of heat, light, or electricity. However, some chemical changes do not require the addition of energy to get them started. The rusting of iron and the tarnishing of silver are examples of such reactions.

As a chemical reaction proceeds, energy is either absorbed or released. For example, the burning of a match releases energy in the form of heat and light. The chemical reaction that occurs in a battery releases electrical energy. On the other hand, when food is cooked, heat energy is absorbed by the chemical changes taking place.

During a chemical reaction called *electrolysis*, electrical energy is absorbed to break down a substance into its component elements. For example, in the electrolysis of water, electricity is used to break apart water molecules, forming hydrogen gas and oxygen gas (see Figure 8-9 on page 133). Table 8-6 on page 133 gives some examples of chemical changes and the forms of energy that are absorbed or released.

Rate of Reactions

As we discussed earlier, milk turns sour when left out of the refrigerator. However, milk that is refrigerated eventually turns sour also, though it takes longer to occur. The chemical reaction of souring, which happens rapidly at room temperature,

Table 8-6. Energy and Chemical Changes

Chemical Changes That Release Energy	Type of Energy Released	Chemical Changes That Absorb Energy	Type of Energy Absorbed
Burning of wood	light, heat	Cooking an egg	heat
Battery powering a flashlight	electricity	Photosynthesis	light
Decomposing of compost	heat	Electrolysis	electricity

occurs much more slowly in a cold refrigerator. In fact, most chemical reactions take place faster at higher temperatures. Frying an egg takes less time on a high flame than on a low flame. An increase in temperature increases the *rate* (speed) of a reaction.

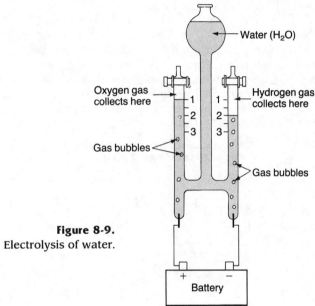

Figure 8-9.
Electrolysis of water.

Another factor that affects the rate of a chemical reaction is the *size* of the reacting particles. In general, the smaller the particles, the faster the reaction. For instance, a log burns more slowly than does an equal quantity of sawdust.

The same factors that influence the rate of a chemical change also affect the rate of many physical changes. Powdered sugar dissolves more rapidly than a cube of sugar does, and sugar dissolves more quickly in hot tea than in iced tea.

PROCESS SKILL: MAKING PREDICTIONS BASED ON RESULTS OF AN INVESTIGATION; DETERMINING A QUANTITATIVE RELATIONSHIP; GRAPHING DATA

A student learned in chemistry class that temperature is a major factor in determining the rate of a chemical reaction. To investigate the effect of temperature on reaction rate, the student decided to time a chemical reaction at several temperatures. The student's results are presented in the table on page 134. Study the table and answer the following questions.

Temperature and Reaction Rate

Trial Number	Temperature	Time for Completion of Reaction
1	20°C	80 seconds
2	30°C	40 seconds
3	40°C	20 seconds

1. The data seem to indicate a trend: for each 10°C increase in temperature, the time needed to complete the reaction was
   (1) doubled
   (2) cut in half
   (3) reduced by 20 seconds
   (4) increased by 20 seconds

2. Assuming that the observed trend remains constant for all temperatures, how long would the reaction take at 50°C?
   (1) 15 seconds   (2) 5 seconds   (3) 10 seconds   (4) 20 seconds

3. Construct a graph by copying the numbered axes provided below onto a separate sheet of graph paper. Then, enter the data from the table, as well as your answer to Question 2, into the graph. To do this, mark a point at the intersection of a temperature line and a time line for the result of each trial. (The result of Trial 1 has been marked as an example to guide you.) Finally, connect the points with a smooth curve.

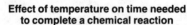

Effect of temperature on time needed to complete a chemical reaction

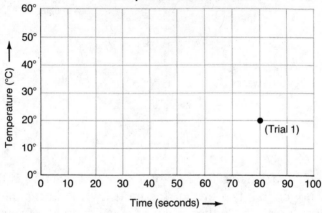

EXERCISE 3

1. Which of the following is only a physical change?
   (1) the souring of milk
   (2) the burning of oil
   (3) the melting of ice
   (4) the rusting of iron

2. Which process involves a chemical change?
   (1) dissolving sugar in water
   (2) boiling water
   (3) freezing water
   (4) electrolysis of water

3. A chemical change always
   (1) forms one or more new substances   (3) releases heat
   (2) absorbs heat   (4) absorbs electricity

4. In making an omelet, which process involves a chemical change?
   (1) melting butter   (2) chopping onions   (3) frying eggs   (4) adding salt

5. A chemist mixed sodium and chlorine but no reaction took place. A probable explanation
   for this is that
   (1) the reaction only releases energy
   (2) the reaction only absorbs energy
   (3) these substances cannot react
   (4) energy must be added to start the reaction

6. During a chemical change,
   (1) energy is always released
   (2) energy is always absorbed
   (3) energy is either absorbed or released
   (4) energy is neither absorbed nor released

7. Melting is a
   (1) chemical change in which energy is absorbed
   (2) chemical change in which energy is released
   (3) physical change in which energy is absorbed
   (4) physical change in which energy is released

8. Hydrogen gas is produced in a chemical reaction between zinc and an acid. Which setup
   would most likely have the fastest reaction rate?

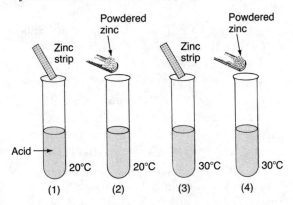

9. As the temperature increases, the rate of a chemical reaction usually
   (1) increases   (2) decreases   (3) remains the same

10. The diagram shows four samples of wood, each weighing one kilogram. Which will most
    likely burn the fastest?

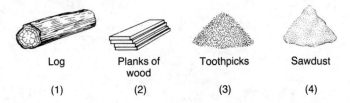

Chapter 9. Energy: Sources and Issues

PART I. SOURCES OF ENERGY

Everything that occurs in the universe involves energy. As you have learned, energy is the capacity to do *work*—the ability to make something move. Heat, light, sound, and electricity are all forms of energy. Humans have learned to describe, explain, and measure energy, and to harness it for their use.

Measuring Energy

To compare the amounts of energy stored in various substances, we need to describe energy with some unit of measurement. The energy in foods and fuels can be measured and compared using a unit called a **calorie**. One calorie is the amount of heat energy needed to raise the temperature of one gram of water by one degree Celsius.

When describing the energy in food, we use the word Calorie spelled with a capital C. This "food Calorie" is equal to 1000 ordinary calories, or 1 *kilocalorie* (*kilo* means one thousand). Calories indicate how much energy you can obtain from various foods. Digested food containing energy that is not needed by the body is stored, usually as fat. When you go on a diet, you count Calories to make sure you don't eat more food than your body needs for energy.

The *rate* at which energy is used (the amount of energy used over a certain time interval) can also be measured. The rate at which electrical energy is used is measured in a unit called the **watt**. We can use watts to compare the rates at which different electrical devices use energy. For instance, a 100-watt light bulb uses twice as much electricity each second as a 50-watt bulb. Table 9-1 lists some common electrical devices and their wattages.

Table 9-1. Some Electrical Devices and Their Wattages

Electrical Device	Wattage
Hair dryer	1200 Watts
Light bulb	100 Watts
Electric shaver	7 Watts
Small air conditioner	860 Watts
Microwave oven	750 Watts
Stereo	240 Watts
Toaster oven	1400 Watts

Energy Consumption

Humanity's consumption of energy is constantly increasing. Our growing populations and economies require more and more energy. *Fuels* are sources of energy. As our demand for energy increases, so does our demand for fuel. We need more fuel to cook meals, heat homes, run industries, and power cars, ships, trains, and airplanes. We also use more fuel to produce electricity.

Electricity has become essential to our society. It is used for many purposes, such as heating and cooling buildings, running machines and appliances, and providing lighting. Electrical energy can be transmitted easily over long distances through conductors such as metal wires. However, electrical energy must itself be produced from other energy sources.

Fossil Fuels

The main energy sources used to produce electricity are the ***fossil fuels***. They are called fossil fuels because they were formed from the remains of plants and animals that lived and died long ago. Over time, these organic remains were changed into energy-rich substances. The most commonly used fossil fuels are oil (also called *petroleum*), coal, and natural gas.

Oil is a sticky black liquid usually found trapped within rock layers deep underground. ***Coal*** is a black rock that occurs in layers, or *seams*, between other rock layers. ***Natural gas*** is commonly found underground with oil deposits. Each of these fossil fuels can be burned to provide energy for the production of electricity. Figure 9-1 shows the relative amounts of oil, coal, and natural gas used to produce electricity in the United States.

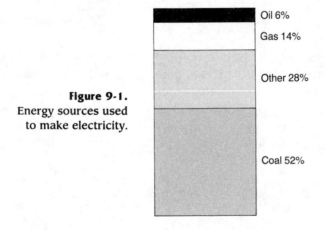

Figure 9-1.
Energy sources used
to make electricity.

Oil 6%

Gas 14%

Other 28%

Coal 52%

Other Uses of Fossil Fuels

Fossil fuels have many uses besides producing electricity. Gasoline is used to power automobiles. Heating fuel is used to heat homes and industries. Both gasoline and heating fuel are obtained from oil and are its main products used for energy. Natural gas is used to heat homes and industries, and for cooking. (Gas stoves use natural gas.) Coal, although mainly used to produce electricity, is also used to provide heat for industrial processes, such as the making of steel.

Fossil fuels are used to make many other important substances. Oil, in particular, has many such uses. Plastics, fertilizers, certain drugs, and synthetic fabrics like nylon and polyester are all products made from petroleum.

PROCESS SKILL: INTERPRETING A GRAPH

A *pictograph* represents numbers, or quantities, by using pictures. In the accompanying diagram, each picture of a barrel represents one million barrels of oil. For example, the category "Industry and Electricity" is represented with three barrels, which means that three million barrels of oil are used each day for these purposes.

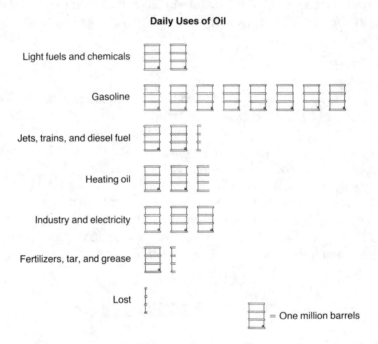

How much oil is used daily for "Jets, trains, and diesel fuel"? Looking at the row of barrels for that category, you see there are two complete barrels, plus part of a third barrel. This part is about one-quarter of a complete barrel, so it represents one-quarter of a million barrels of oil. This makes the total daily use of oil for "Jets, trains, and diesel fuel" equal to 2¼ million barrels. Use the diagram to help you answer the following questions.

1. Which products use the least amount of oil each day?
   (1) light fuels and chemicals   (3) fertilizers, tar, and grease
   (2) heating oil   (4) gasoline

2. About how much oil is lost every day?
   (1) half a million barrels
   (2) less than one-quarter of a million barrels
   (3) one million barrels
   (4) more than one-quarter of a million barrels

Other Sources of Energy

While most of the energy we consume comes from fossil fuels, there are several other energy sources currently in use. The most important of these are *hydroelectric energy* and *nuclear energy*.

(1) **Hydroelectric energy** is electricity produced by the power of flowing water. Water in motion has kinetic energy, which can be transformed into electrical energy. When water flows steeply downhill at a waterfall or a dam on a river, it can be used to operate a *generator*, which produces electricity. A generator contains a *turbine*, a device similar to a paddle wheel on an old-fashioned steamboat (Figure 9-2). The moving water turns the turbine, which spins a coil of wire inside an electromagnet in the generator. This creates an electrical current.

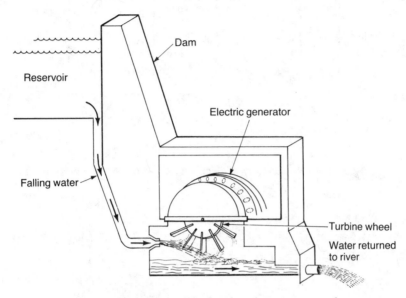

Figure 9-2. Production of hydroelectric energy at a dam.

Hydroelectric energy is usually clean and inexpensive. Niagara Falls is a major source of electrical energy in New York State. Figure 9-3 compares the energy sources used by electric companies in New York with those used in the entire nation. Notice that New York uses a greater percentage of hydroelectric energy than does the nation as a whole, while the percentage of coal used in New York is much smaller.

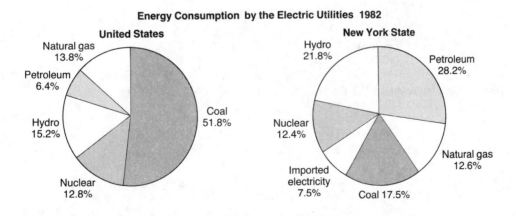

Figure 9-3. Comparison of energy sources used for electricity in New York State with sources used in the entire United States.

(2) **Nuclear energy** is the energy stored in the nucleus of an atom. When this energy is released, it creates heat, which can be used to produce electricity. Nuclear power plants use **uranium** (a radioactive element found in certain rocks) as their fuel source. Uranium atoms are naturally unstable, and can be readily split apart to release

Nuclear power plant

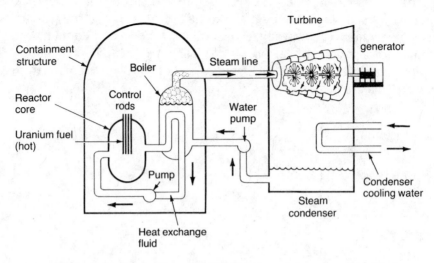

Figure 9-4. Production of electricity at a nuclear power plant.

heat energy. The heat created by the uranium fuel in a nuclear reactor is used to boil water, thereby producing steam. The steam turns turbines that generate electricity. This is illustrated in Figure 9-4.

EXERCISE 1

1. Which is *not* a form of energy?
   (1) heat   (2) electricity   (3) gasoline   (4) light

2. The energy obtained from foods is generally measured in
   (1) Calories   (2) watts   (3) volts   (4) degrees

3. A unit that measures the rate at which electrical energy is used is a
   (1) volt   (2) watt   (3) calorie   (4) degree

4. Which is *not* a fossil fuel?
   (1) uranium   (2) coal   (3) oil   (4) natural gas

5. Plastics, fertilizers, and synthetic fabrics are products commonly made from
   (1) coal   (2) oil   (3) gasoline   (4) uranium

6. Fossil fuels were formed from
   (1) rocks and minerals
   (2) uranium deposits
   (3) remains of dead plants and animals
   (4) moving water

7. Higher gasoline prices would most likely result from a shortage of
   (1) coal   (2) oil   (3) natural gas   (4) heating fuel

Base your answers to questions 8–10 on the graphs below, which compare energy used in the entire United States with energy used in New York State.

**Energy Use by Source
In the United States and New York State**

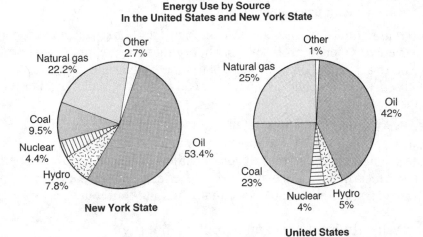

New York State

United States

8. More than half the energy used in New York State comes from
   (1) natural gas
   (2) coal
   (3) hydroelectric power
   (4) oil

9. Compared to the entire United States, New York State uses
   (1) a larger percentage of oil and a larger percentage of coal
   (2) a smaller percentage of oil and a larger percentage of coal
   (3) a smaller percentage of oil and a smaller percentage of coal
   (4) a larger percentage of oil and a smaller percentage of coal

10. The total percentage of energy used in the United States that comes from fossil fuels is
    (1) 10% (2) 48% (3) 65% (4) 90%

PART II. PROBLEMS WITH ENERGY SOURCES

Problems With Fossil Fuels

In the United States, most of the demand for energy is met by the fossil fuels: oil, coal, and natural gas. However, the burning of fossil fuels creates air pollution. When fossil fuels burn, chemicals are released into the air that pose dangers to living things and their environment. This is especially true of coal.

1. Coal. The supply of coal found in the United States is much greater than that of oil or natural gas. Because of its abundance, coal is relatively inexpensive. However, there are serious environmental and health problems involved with its use. The burning of coal contributes greatly to air pollution. Smoke from coal-burning power plants is the main cause of *acid rain*, which is harmful to the ecology of lakes and forests.

In addition, coal mining is dangerous for people who work in the mines. Breathing air that contains coal dust is unhealthy and can lead to *black lung disease*. Certain coal-mining techniques are also damaging to the environment. Sometimes large areas of land are dug up to reach the coal. This practice, called *strip mining*, destroys topsoil and scars the landscape.

Mining companies are now required by law to restore the land they have damaged. Advances in technology have made coal mining safer and reduced the amount of pollution caused by burning coal. Nevertheless, these measures have only begun to solve the problems with using coal as a fuel.

2. Oil and Natural Gas. Although oil and natural gas burn cleaner than coal, they cause other environmental problems. Offshore drilling for oil and transporting oil by ship can lead to accidental oil spills that kill marine wildlife and cause severe pollution of land and sea. Pipelines built to transport oil and gas over land may alter the ecology of areas they cross.

3. Global Warming. The burning of any fossil fuel produces carbon dioxide. Carbon dioxide traps heat in Earth's atmosphere much as the glass of a greenhouse traps heat. Some scientists fear that the buildup of carbon dioxide in the atmosphere caused by using fossil fuels may produce a **greenhouse effect**, leading to a warming of Earth's climate. Such global warming could have many harmful consequences for life on Earth.

Problems With Hydroelectric and Nuclear Energy

The production of electricity using moving water or nuclear reactions is generally much "cleaner" than energy production with fossil fuels. This is because hydroelectric and nuclear power do not involve burning anything. However, even these "clean" energy sources have environmental costs.

1. Hydroelectric Energy. Building a dam on a river to produce hydroelectric power changes the surrounding area, as you can see in Figure 9-5. The area upriver is flooded, creating a large lake over land that may have once provided a habitat for wildlife or been used for farming. The area downriver from the dam receives a diminished flow of water. These changes greatly affect the ecology of the area around a dam.

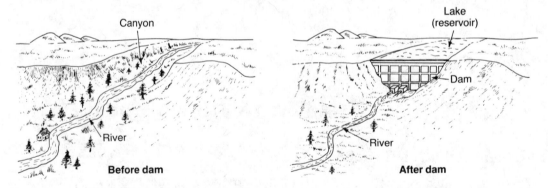

Figure 9-5. Before and after construction of a dam on a river.

2. Nuclear Energy. Although nuclear power plants do not cause air pollution, they use water from nearby lakes or rivers to cool their nuclear reactors. The water is then returned to the environment several degrees warmer. This increase in the temperature of the environment, called **thermal pollution**, can be harmful to organisms living in the water.

An even more serious problem with nuclear power is how to safely dispose of the used-up uranium fuel, known as **nuclear waste**. This poisonous, radioactive material must be stored where it will never leak into the environment. Most people do not want nuclear waste stored in, or even transported through, their communities. Disposal of nuclear waste is a difficult problem, for which no adequate solution has yet been found.

EXERCISE 2

1. Most of the energy consumed in the United States comes from
   (1) nuclear energy   (2) fossil fuels   (3) moving water   (4) solar energy

2. Which fossil fuel is most abundant in the United States?
   (1) oil   (2) coal   (3) natural gas   (4) uranium

3. Killing of marine wildlife and pollution of shorelines result from accidents involved in
   (1) coal mining   (3) building dams
   (2) storing nuclear waste   (4) transporting oil by ship

4. Acid rain is mainly caused by
   (1) burning natural gas   (3) drilling for oil
   (2) nuclear reactors   (4) burning coal

5. The threat of global warming is mainly associated with
   (1) radioactive wastes   (3) burning of fossil fuels
   (2) use of hydroelectric energy   (4) thermal pollution of lakes and rivers

6. Which energy source produces the most air pollution?
   (1) coal   (2) uranium   (3) moving water   (4) natural gas

7. Thermal pollution from nuclear power plants involves
   (1) release of toxic chemicals into the environment
   (2) storage of nuclear wastes
   (3) an increase in the temperature of the environment
   (4) a decrease in the temperature of the environment

8. The safe storage of hazardous wastes is a problem involved mainly with
   (1) transporting oil   (3) hydroelectric power
   (2) strip mining   (4) nuclear power

9. A problem with dams and pipelines is that they
   (1) contribute to air pollution
   (2) produce toxic wastes
   (3) cause thermal pollution
   (4) alter the ecology of surrounding areas

10. The energy source being used to make electricity in the diagram below is
    (1) nuclear power
    (2) moving water
    (3) a fossil fuel
    (4) wind

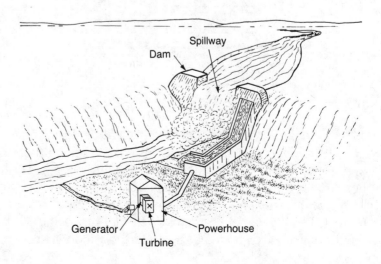

PART III. ENERGY FOR THE FUTURE

Energy Conservation

Most energy resources do not exist in unlimited amounts. To guarantee an adequate supply of energy for the future, we must practice conservation. *Conservation* is the saving of natural resources through wise use. This means using resources more efficiently, and eliminating unnecessary waste.

1. High-Efficiency Appliances. We can contribute to conservation efforts by purchasing high-efficiency appliances. These appliances consume less energy than less efficient appliances, while doing the same job. For instance, a car that can travel 30 miles on a gallon of gasoline is more efficient than a car that gets only 15 miles per gallon. An air conditioner with a high "energy efficiency rating" uses less electricity than one with a low rating, but it cools a room just as well.

Although a high-efficiency appliance may cost more to buy, it costs less to use. This means that it will save money in the long run, while helping to conserve energy. Many appliances carry an "Energy Efficiency Rating" label that shows how they compare with other models. Figure 9-6 shows an example of such a label.

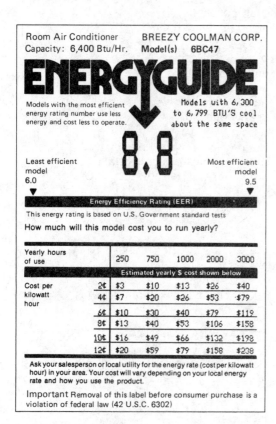

Figure 9-6.
An Energy Efficiency Rating label.

2. Insulated Buildings. Energy use at home and at work can be reduced through improved *insulation*. A well-insulated building prevents heat loss in winter and keeps heat out in summer. These benefits can be achieved by constructing walls in two layers, with insulating material in between. Cracks around doors and windows can be sealed with weather stripping for further insulation. With these improvements, less energy is needed to maintain a comfortable indoor temperature year-round.

3. Recycling. This practice also helps conserve energy resources. The graph in Figure 9-7 shows that making bottles and cans from recycled materials consumes less energy than does making those products from raw materials.

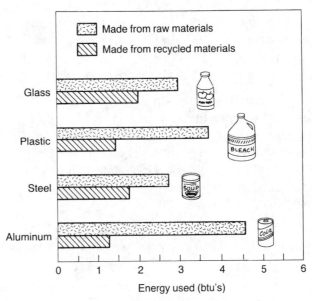

Figure 9-7.
Recycling discarded containers uses less energy than making containers from raw materials does.

Renewable and Nonrenewable Resources

Earth's supply of fossil fuels is rapidly being used up. We continually remove these resources from the earth, but we cannot replace them. Nature does not create new deposits of oil, coal, and natural gas within the time span of human history. For this reason, fossil fuels are considered *nonrenewable resources*. Uranium is also a nonrenewable resource.

Renewable resources are those that can be replenished by nature within a relatively short time span. Moving water, wind, plants, and sunshine do not run out as we use them, because they are constantly being replaced by natural processes. Table 9-2 lists some renewable and nonrenewable resources.

Table 9-2. Energy Resources

Renewable	Nonrenewable
Hydroelectric	Oil
Solar	Coal
Wood	Natural gas
Wind	Nuclear

Using Renewable Resources

Even if we practice conservation, our supply of nonrenewable energy resources may not be sufficient to meet the energy demands of the future. Renewable resources offer alternatives to fossil fuels and radioactive minerals. Unlike nonrenewable resources such as oil, coal, and uranium, renewable energy resources cannot run out. They also cause fewer environmental problems than fossil fuels and nuclear energy do. For these reasons, scientists and engineers are seeking more and better ways to use renewable resources for our growing energy needs.

As you know, moving water can be used to run generators and produce electricity. The natural water cycle of evaporation, condensation, and precipitation renews the water supply that feeds the rivers used for this purpose. However, not all areas have rivers suitable for producing hydroelectric energy.

The wind can be used to generate electricity by turning the blades of a *wind turbine* (see Figure 9-8 on the facing page). In windy areas, wind turbines can provide safe, clean electricity. But the wind is not as constant and reliable as a flowing river. When there is only a slight wind or the air is calm, little or no electricity is produced.

Plant matter and animal wastes can be burned to produce heat, or they can be changed to other fuels. For instance, decaying plant matter and animal wastes produce *methane*, the main component of natural gas. Methane produced in this way is a renewable resource, unlike natural gas found underground. At present, however, converting these materials into fuel on a large scale is too expensive to be practical. Unfortunately, *every* energy source has both advantages and disadvantages, as outlined in Table 9-3.

Table 9-3. Advantages and Disadvantages of Energy Sources

Energy Source	Advantages	Disadvantages
Oil	Efficient; can be converted into different types of fuel	Causes air pollution; risk of spills while drilling or transporting; limited reserves in U.S.; nonrenewable
Natural gas	Available in U.S.; clean	Difficult to store and transport; mostly nonrenewable
Coal	Abundant in U.S.; inexpensive	Causes air pollution and acid rain; mining practices harmful to miners' health and destructive to environment
Nuclear	Highly efficient; does not cause air pollution; inexpensive; uranium fuel abundant in U.S.	Causes thermal pollution; creates radioactive waste; risk of accidents releasing radioactivity into environment; uranium mining harmful to miners' health
Hydroelectric	Does not cause air pollution; inexpensive; renewable	Not available in all areas; affects local ecology
Wind	Does not cause pollution; clean; inexpensive; renewable	Not practical for large-scale power generation; not always reliable (winds not constant)
Solar	Does not cause pollution; clean; renewable	Expensive to convert into usable form; not always reliable (depends on sunny weather)
Plant matter and animal wastes	Renewable	Expensive to convert into usable form; inefficient

Solar Energy

The primary source of energy on Earth is the sun. Energy from the sun is called **solar energy.** The energy in fossil fuels came originally from sunlight absorbed by

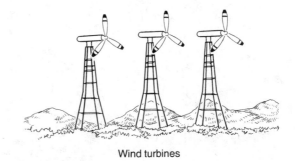

Figure 9-8. Wind turbines use the energy of the wind to produce electricity.

Wind turbines

plants during photosynthesis, millions of years ago. The moving water used for hydroelectric energy is replenished by the water cycle, which is powered by the sun's energy. And wind, which can be used to make electricity, is caused by the sun's heating of the atmosphere.

People have found ways to use the sun's energy directly to provide heat and hot water for homes, offices, and factories. For example, a device called a *solar collector* absorbs solar energy and converts it into heat energy. The heat is transferred to water circulating through the collector. This hot water can be used to run a home heating system, as shown in Figure 9-9.

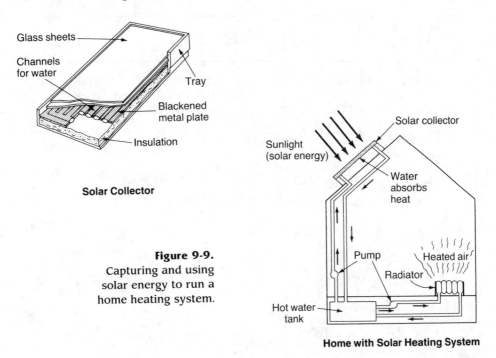

Solar Collector

Figure 9-9. Capturing and using solar energy to run a home heating system.

Home with Solar Heating System

People have also developed ways to transform solar energy into electrical energy. For instance, a *solar cell* is a device that converts light directly into electricity. Some calculators and light meters in cameras use solar cells. However, to generate large amounts of electricity this way requires a huge number of these cells, which are very expensive.

Electricity can also be produced by heating water with solar energy. Water is heated to a boil by using mirrors to focus and concentrate sunlight (see Figure 9-10 on page 148). The boiling water changes to steam, which turns turbines to produce electricity. This method is more economical than using solar cells to make electricity, but it is still more expensive than using fossil fuels. However, as fossil fuels become more scarce, this situation may change.

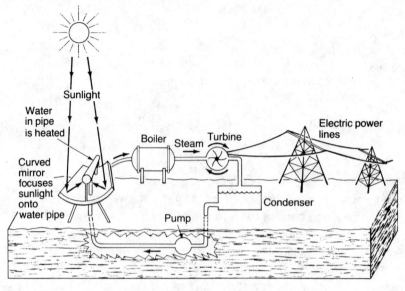

Figure 9-10. Solar energy changes water into steam, which turns turbines to produce electricity.

EXERCISE 3

1. Energy resources can be conserved by all of the following means *except*
   (1) better insulation
   (2) using high-efficiency appliances
   (3) increased mining of coal
   (4) increased use of solar energy

2. Which of the following is a renewable energy resource?
   (1) moving water   (2) uranium   (3) coal   (4) oil

3. The primary source of most energy on Earth is
   (1) moving water   (2) the sun   (3) coal   (4) wind

4. In the diagram below,

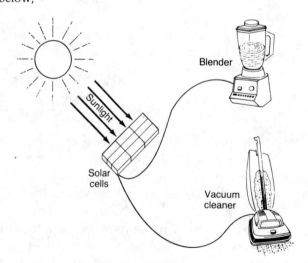

   (1) electrical energy is being changed into solar energy
   (2) solar energy is being changed into electrical energy
   (3) chemical energy is being changed into solar energy
   (4) solar energy is being changed into chemical energy

5. Because natural gas taken from underground is not quickly replaced by nature, it is considered a
   (1) renewable resource
   (2) nonrenewable resource
   (3) pollutant
   (4) solar energy source

6. *Gasohol* is a combination of gasoline and alcohol. Gasoline comes from oil, and alcohol is made from plant matter. Gasohol was developed to decrease our use of oil. However, gasohol costs more than ordinary gasoline and is therefore not commonly used.

   The above paragraph suggests that
   (1) gasohol will soon replace gasoline
   (2) gasohol causes less pollution than gasoline
   (3) renewable resources are less efficient than nonrenewable resources
   (4) alternative energy sources may have both advantages and disadvantages

Questions 7–10 are based on the graph below, which shows the monthly gas and electric bills of a family in New York State for one year.

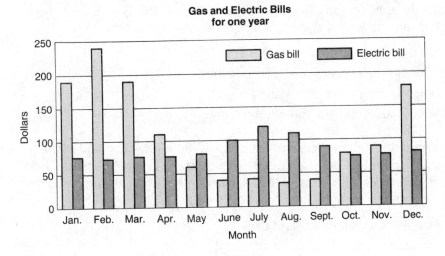

7. For which month was the electric bill highest?
   (1) February (2) June (3) July (4) December

8. For which month were the gas and electric bills about the same?
   (1) March (2) June (3) October (4) December

9. These bills are probably for a home that is heated by
   (1) oil (2) electricity (3) wood (4) gas

10. What would most likely result from improved insulation of this home?
    (1) Gas bills would decrease, and electric bills would increase.
    (2) Gas bills would increase, and electric bills would decrease.
    (3) Both gas and electric bills would decrease.
    (4) Both gas and electric bills would increase.

Chapter 10. Science, Technology, and Society

PART I. RELATIONSHIP OF SCIENCE AND TECHNOLOGY

Science and Technology

Science and technology affect the lives of people all over the world. *Science* is the process of asking questions and seeking their answers to gain an understanding of the natural world. By providing insight into the workings of nature, science helps us predict the outcome of physical events.

Some questions that science attempts to answer include:

- What is the nature of matter?
- How did the universe come into being?
- How did life evolve on Earth?
- What causes Earth's climate to change?

Technology is the process of using scientific knowledge and other resources to develop new products and processes. These products and processes help solve the problems and meet the needs of the individual or society. Some problems that technology attempts to solve include how to:

- increase the gas mileage of cars
- increase the productivity of farmland
- control industrial pollution
- improve satellite telecommunications

While the emphasis in science is on gaining knowledge of the natural world, the emphasis in technology is on finding practical ways to apply that knowledge to solve problems.

There are three major fields of science: life science, earth science, and physical science. Each of these fields contains a number of more specific sciences (Figure 10-1).

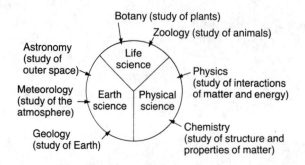

Figure 10-1.
The major fields of science.

Biologists, chemists, physicists, astronomers, and geologists are some types of scientists.

Engineers, computer programmers, and medical technicians are examples of workers in the fields of technology. Resources are essential for the progress of technology. These resources include materials (both natural and synthetic), money, people, energy, information, and time.

Science and Technology Advance Each Other

Science and technology frequently help to advance each other. Scientific discoveries often lead to the development of new or better technological devices and processes. These technologies may, in turn, lead to new discoveries or to a better understanding of scientific principles.

For instance, scientists discovered various properties of light, such as how light is bent when it passes through different types of lenses. This knowledge led to the invention of the telescope and the microscope. Using these technological devices, scientists have made many more discoveries about the natural world.

Every technological device or process is based in some way on scientific principles, as the examples in Table 10-1 suggest.

Table 10-1. Relation of Scientific Principles and Technology

Scientific Principle	*Technological Device or Process*
Cold temperatures kill or reduce growth of microorganisms	Refrigerators and freezers
Sunlight contains energy	Solar heating systems and solar cells
Lenses bend light rays	Telescopes, microscopes, and binoculars
Splitting atoms of radioactive elements produces heat	Nuclear power plants
Every action produces an equal and opposite reaction	Rocket engines and jet engines

In fact, much technology involves knowledge from more than one field of science. For example, the artificial heart shown in Figure 10-2 involves knowledge from both the life sciences (the structure of the human heart) and the physical sciences (the mechanical principles of how the heart works).

Figure 10-2.
The artificial heart.

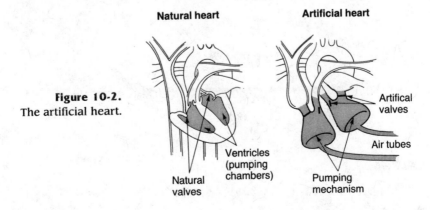

Sometimes technology is developed without full knowledge of the scientific principles on which it is based. This technology may then help lead to the discovery of those principles. For instance, the airplane was invented before the principles of flight were fully understood. Instead, the airplane's development helped bring about an understanding of those principles.

Technology as a System

Some technological processes and devices can be viewed as systems. A *system* is a group of related elements or parts that work together for a common purpose. The parts of a system act in a series of steps, consisting of *input, comparison* and *control, processing, output,* and *feedback.*

A home heating system with a furnace and a thermostat can be viewed in terms of these steps (Figure 10-3). Setting the thermostat to the desired temperature is the *input.* The thermostat *compares* the actual room temperature to the set temperature and *controls* the furnace, turning it on if the room's temperature is too low. The burning of fuel in the furnace is the *process* that produces the *output*—heat. The changing room temperature provides *feedback* to the thermostat, which turns the furnace off when the desired temperature is reached. In this way, the system maintains a constant indoor temperature.

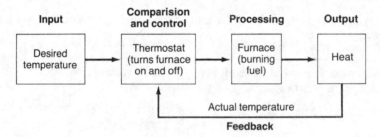

Figure 10-3. The steps of a heating system.

A thermostat is a fairly simple device. A complex technological device or process is made up of a number of simpler devices or processes that work together as a system. Each of these simple devices or processes is a *subsystem.* An automobile is a complex system made up of many subsystems (Figure 10-4). One subsystem is the cooling system, which prevents the car's engine from overheating. Other subsystems in a car include the braking system, the transmission, the steering system, and the engine. All of these subsystems must work together for the car to function properly.

Further examples of complex technological systems include a telephone network, a factory assembly line, and a spacecraft like the space shuttle.

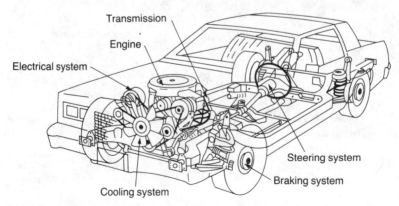

Figure 10-4. An automobile is a system with many subsystems.

Technology in Use

Everyone interacts with the products of technology. You do so when you wear clothing, sleep on a bed, watch television, eat with a knife and fork, ride a school bus—in almost everything you do.

People use technology for a number of reasons (Figure 10-5). To extend or improve our abilities, we use radios and telephones, calculators and computers, binoculars and telescopes, and other devices. Machines and appliances help us do work that requires more than human strength, and at faster speeds than are humanly possible. To overcome physical disabilities, people use devices like eyeglasses, hearing aids, and heart pacemakers.

Telescope   Sewing machine   Wheelchair

Figure 10-5. Some technological devices are used to extend human abilities or overcome handicaps.

Many products of technology are used to change our environment. We use electric lights so that our activities can continue after nightfall. Heaters and air conditioners maintain a comfortable indoor environment throughout the year. We build dams to store water for our cities, generate hydroelectric energy, and create lakes for recreation.

Every technological process or device affects the environment in some way. Some of these effects may be harmful. Light bulbs and many other appliances require electricity. The production of electricity may use precious natural resources or cause pollution of air and water. Cars, boats, and airplanes also consume energy resources and cause pollution. Even eating utensils have an impact on the environment, since energy and mineral resources are used to make them, and they may someday be thrown out as trash. However, technology can also be used to protect the environment, as with sewage treatment plants and pollution-control devices in cars and factories.

New technology often builds upon and improves past technology. For example, contact lenses were developed from eyeglasses, jet airplanes were developed from propeller-driven airplanes, and battery-powered watches were developed from watches driven by winding a spring.

A technological device or process may become outdated, or *obsolete*, when a new device or process is developed that does its job better. Slide rules, fountain pens, and iceboxes are obsolete devices. Slide rules have largely been replaced by pocket calculators, fountain pens by ballpoint pens, and iceboxes by freezers and refrigerators. Today, phonograph records are rapidly being replaced by compact discs and cassette tapes, so that records may become obsolete.

Sometimes, obsolete technology is improved or updated and becomes useful again. For instance, windmills were once used to grind wheat into flour. When machinery was invented that did this job more efficiently, windmills became obsolete. However, modern *wind turbines*, which are used to generate electricity in some regions, operate in practically the same way that traditional windmills did long ago (see Figure 10-6 on page 154).

PROCESS SKILL: GRAPHING DATA

The table below lists the numbers of different word-producing machines used in a newsroom from 1955 to 1990, at five-year intervals. These numbers show how the usage of manual typewriters, electric typewriters, and word processors has changed over time.

Numbers of Different Word-Producing Machines in a Newsroom, 1955–1990

Year	Manual Typewriters	Electric Typewriters	Word Processors
1955	45	0	–
1960	35	10	–
1965	25	20	–
1970	10	40	–
1975	0	50	0
1980	–	40	10
1985	–	15	40
1990	–	5	50

The relationships among the three sets of data may not be immediately clear from the table. However, if the same data are presented in a graph, these relationships become much easier to interpret.

The graph on the facing page shows only the data for usage of manual typewriters (represented by triangles). Copy the graph on a separate sheet of graph paper. Then enter the data for electric typewriters, using circles, and the data for word processors, using squares. Draw lines connecting circles to circles, and squares to squares. (It may be helpful to use a different color pencil for each set of data.) Then answer the following questions.

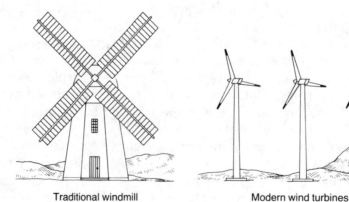

Traditional windmill   Modern wind turbines

Figure 10-6. The modern wind turbine is an updated version of the traditional windmill.

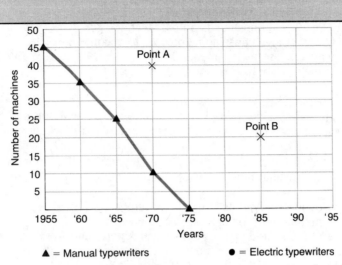

▲ = Manual typewriters   ● = Electric typewriters

■ = Word processors

1. Point *A* on the graph represents the number of
   (1) manual typewriters in use in 1970
   (2) electric typewriters in use in 1980
   (3) electric typewriters in use in 1970
   (4) word processors in use in 1970

2. Point *B* represents the number of
   (1) electric typewriters in use in 1985
   (2) word processors in use in 1980
   (3) word processors in use in 1985
   (4) none of the above

3. If the current trend for electric typewriters continues, the number of electric typewriters in use in 1995 will be
   (1) less than in 1990   (3) the same as in 1990
   (2) more than in 1990   (4) impossible to determine

EXERCISE 1

1. Science is the process of
   (1) solving the problems and meeting the needs of individuals and society
   (2) gaining an understanding of the natural world
   (3) developing new products and processes
   (4) all of the above

2. The process of using scientific knowledge to develop new products or processes is called
   (1) science   (2) industry   (3) technology   (4) renewing resources

3. Using scientific knowledge about magnetism and electricity to build an electromagnet is an example of
   (1) a scientific discovery   (3) predicting future physical events
   (2) a technological development   (4) observing the natural world

4. Using scientific knowledge, engineers build a space probe and send it to the planet Jupiter. The probe sends data about Jupiter back to Earth, adding to our scientific knowledge. Which statement does this best demonstrate?
   (1) New technology sometimes builds on past technology.
   (2) Advances in technology cause some devices to become obsolete.
   (3) Technology affects our environment.
   (4) Science and technology help to advance each other.

5. An airplane is an example of
   (1) a system and many subsystems
   (2) several systems and many subsystems
   (3) a system without any subsystems
   (4) many unrelated systems

6. Which of these technological devices affects the environment in some way?
   (1) coffeepot   (2) washing machine   (3) air conditioner   (4) all of these

7. Which of the following statements is most correct?
   (1) Few people interact with the products of technology.
   (2) Only people in the United States interact with the products of technology.
   (3) Only adults interact with the products of technology.
   (4) Everyone interacts with the products of technology.

8. In this air-conditioning system, the *output* is the
   (1) desired temperature
   (2) cool air
   (3) actual temperature
   (4) thermostat

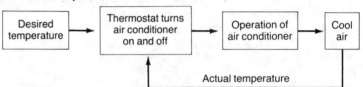

9. The chart below lists three general uses of technology. Examples of uses 1 and 2 are given in the chart. Which three examples would best fit in the third column?

Uses of Technology

1. *Affect Our Environment*	2. *Overcome Disabilities*	3. *Extend Our Abilities*
Air conditioner	Hearing aid	
Space heater	Wheelchair	
Dam on a river	Eyeglasses	

   (1) heart pacemaker, furnace, traffic light
   (2) binoculars, telephone, calculator
   (3) hammer, furnace, videocassette recorder
   (4) airplane, plumbing, coal mining

10. The difference between the two heating systems shown below is that system *A* is lacking
    (1) input
    (2) output
    (3) feedback
    (4) processing

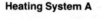

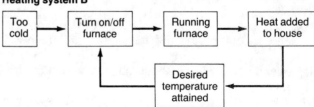

PART II. INTERACTION OF SCIENCE, TECHNOLOGY, AND SOCIETY

Effects of Science and Technology on Society

Science, technology, and society are constantly interacting (Figure 10-7). Often, a change in one of these areas will affect the other two. For example, scientific discoveries about the structure of matter led to many technological developments, including the production of microprocessors on tiny silicon chips. These "microchips" made possible many new products that have affected society by improving health care, communications, and transportation.

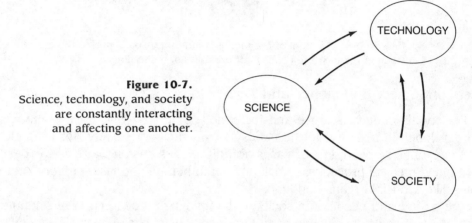

Figure 10-7.
Science, technology, and society are constantly interacting and affecting one another.

Our culture, economy, and social systems are often affected by developments in science and technology. During the 1800s, the United States was transformed from a mainly agricultural society to a highly industrialized society. This period of cultural, economic, and social change was caused by the development of industrial machinery and new ways to power it.

Many products have been made available or improved by science and technology. These products have raised our standard of living and changed our life-styles. Advances in science and technology have helped to decrease our working hours, improve health care, and create many work-saving home appliances. As a result, we have more leisure time, and we live longer, healthier lives.

Science and technology have helped create new businesses and industries. The motion picture industry and the music recording industry did not exist before technology was developed to record sounds and images on tape and film. Record stores and video stores also owe their existence to these technologies. On the other hand, technological developments have eliminated some industries. For example, the ice-cutting industry was made obsolete by the development of refrigerators and freezers.

Developments in science and technology also affect career choices and job opportunities. They have eliminated some jobs, modified other jobs, and created new jobs. For example, some telephone operators have been replaced by computers, many typists have had to learn to use word processors, and there are many new jobs for computer analysts.

Advances in science and technology have helped solve some of society's problems. For example, the development of vaccines has virtually eliminated certain diseases, such as smallpox. However, science and technology alone cannot solve all of society's problems. World hunger, the threat of nuclear war, and overpopulation are problems that cannot simply be solved by technological advances. They also require the willingness and cooperation of people everywhere to find solutions.

While science and technology have solved many problems, they have also created problems. Pollution of the environment and disposal of garbage and hazardous waste are problems caused, in part, by science and technology (Figure 10-8). Solving such problems requires the help of people working in government, industry, science, and technology.

Figure 10-8. Disposal of garbage is a mounting social problem. Much of the garbage consists of products of technology.

Effects of Society on Science and Technology

Society also affects science and technology in many ways. New technology is often developed in response to the needs of individuals or society. For example, the need to help people overcome diseases and disabilities has encouraged the development of new medical procedures, such as chemotherapy and laser surgery, and new devices, like artificial organs and limbs.

The attitudes of people in a society may influence the direction of scientific research and technological development. In our society, public opinion has encouraged research to find a cure for AIDS. In contrast, public attitudes have largely discouraged the use of animals to test the safety of new cosmetic products.

Acceptance and use of an existing technology can also depend on people's attitudes. An example is nuclear energy. Most people agree that nuclear energy has both benefits and drawbacks. However, people disagree about whether or not its benefits outweigh its dangers. Public attitudes against nuclear energy have led some countries to ban its use. Other countries, on the other hand, generate most of their electricity with nuclear energy. Public opinion will undoubtedly influence the future of nuclear energy (Figure 10-9).

Society may also affect science and technology by providing money for research and development. Our government finances space exploration, medical research, de-

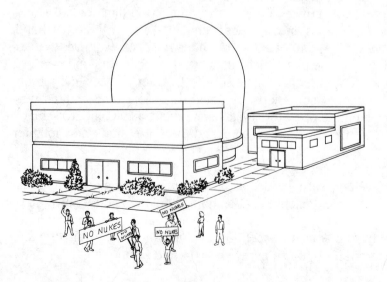

Figure 10-9. People's attitudes toward nuclear power will affect the acceptance and use of this technology.

velopment of military technology, and study of the environment. In addition, some industries provide funding for research and development in their areas of interest. Many oil companies, for example, fund research to find new ways to locate and recover oil.

Global Effects of Technology

Technology used in one country may have an international or global impact. For instance, in 1985, an accident at the Chernobyl nuclear power plant in the Soviet Union released radiation that affected several neighboring countries. The radiation contaminated livestock, crops, and water. Another example is acid rain. Industries in the Midwest create air pollution that drifts eastward with the prevailing winds. This causes acid rain to fall in New York, New England, and parts of Canada.

A possible global effect of technology is the destruction of Earth's *ozone layer*. Certain chemicals used in refrigerators, air conditioners, and spray cans are destroying the layer of ozone gas that exists high in the atmosphere. The ozone layer screens out much of the sun's dangerous ultraviolet radiation. Its destruction could cause an increase in skin cancer, and may have harmful effects on Earth's ecology.

On the positive side, people around the world interact more frequently because of technological advances. Air travel enables people to reach destinations all over the world within a day or less. Communications satellites let us make phone calls to people on other continents. We can also view distant events on television with the help of satellites (Figure 10-10). Technology helps inform us immediately of natural disasters like earthquakes and hurricanes, and it provides the means of sending aid to victims of these disasters.

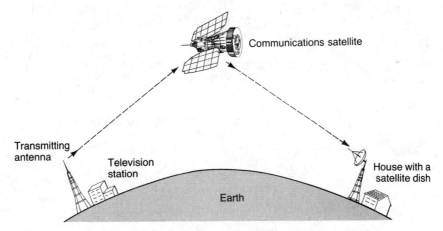

Figure 10-10. Communications satellites enable us to view distant events on television.

EXERCISE 2

1. Society affects technology by
   (1) having problems that need to be solved
   (2) providing funds for research and development
   (3) its attitudes toward new research or products
   (4) all of the above

2. An example of a job that has been created by recent developments in technology is
   (1) farmer   (2) astronaut   (3) schoolteacher   (4) postal worker

3. The problem of environmental pollution can only be solved with the help of people working in the areas of
   (1) science and technology
   (2) government and industry
   (3) technology and industry
   (4) science, technology, government, and industry

4. Developments in microelectronics and computer science are changing the United States from a largely industrial nation into one more dependent on information services. This is an example of
   (1) science helping technology to advance
   (2) society affecting science and technology
   (3) science and technology affecting society
   (4) science and technology solving society's problems

5. In Europe, many people use *irradiated* milk (milk subjected to radiation that kills micro-organisms). Containers of such milk can be stored at room temperature for a long time if unopened. In the United States, however, many people regard irradiated milk with suspicion, so its use has not become popular.

   The above paragraph illustrates that
   (1) technology has affected society by raising our standard of living
   (2) people's attitudes can affect acceptance and use of technological devices or processes
   (3) products of technology may have an international impact
   (4) technology has caused some industries to become obsolete

6. Carbon dioxide released into the atmosphere by industry and cars may cause a warming of Earth's climate, called the greenhouse effect. This is an example of
   (1) a problem solved by technology
   (2) people's attitudes affecting the use of technology
   (3) new industries being created by technology
   (4) a global impact of technology

7. The table below shows how the percentage of the work force in three fields has changed over the years.

Percent of Work Force in Three Fields from 1800 to 1990

Year	Agriculture	Information	Industry
1800	75%	5%	20%
1850	60%	5%	35%
1900	40%	5%	55%
1950	10%	20%	70%
1990	5%	75%	20%

   From 1800 to 1990, the percent of the work force in industry
   (1) increased steadily
   (2) decreased steadily
   (3) first increased and then decreased
   (4) first decreased and then increased

8. In the past, lumberjacks chopped down trees using axes and handsaws. Today, they mostly use motorized chain saws. This example demonstrates that
   (1) technology has created new jobs
   (2) technology has modified some jobs
   (3) technology has eliminated some jobs

9. The graph below shows how average life expectancy has changed over time.

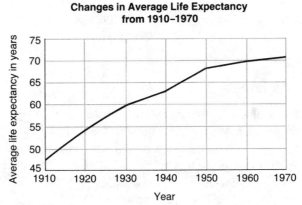

**Changes in Average Life Expectancy
from 1910–1970**

This change is most likely a result of
(1) advances in medical technology
(2) harmful effects of technology on the environment
(3) advances in educational technology
(4) advances in communication technology

10. Which answer is the best example of the relationship shown in the diagram below?

(1) principles of electricity → development of electrical appliances → more leisure time for people
(2) effect of cold on microorganisms → people eat fresher, healthier food → development of refrigerators and freezers
(3) principles of nuclear energy → development of nuclear power plants → development of nuclear weapons
(4) development of radio and television → principles of electromagnetic waves → home entertainment for people

PART III. MAKING DECISIONS ABOUT TECHNOLOGY

Technology Increases Our Choices

People have more choices in their everyday lives because of the products of technology. For example, cable television and videocassette recorders have increased our choices in home entertainment. Figure 10-11 shows some leisure activities that are outgrowths of technology.

Waterskiing   Bicycle riding   Scuba diving

Figure 10-11. Many leisure-time activities are outgrowths of technology.

Technology has given us more forms of transportation to choose from. For instance, people may travel from New York City to Boston by car, bus, train, or airplane. Shoppers may choose from a wide selection of home appliances produced by technology. Coffee makers, microwave ovens, dishwashers, and vacuum cleaners are just a few of these products.

Assessing Technology

Every technological process or device has advantages and disadvantages associated with its use, providing both **benefits** and **burdens** for people and the environment. For instance, the automobile has given people greater mobility and contributed to our nation's prosperity. However, cars and trucks contribute to air pollution and lead to deaths and injuries in traffic accidents. Table 10-2 lists benefits and burdens of some technological devices and processes.

Table 10-2. Benefits and Burdens of Technology

Technological Process or Device	Benefits	Burdens
Nuclear energy	Additional clean electricity	Risk of accidents; radioactive wastes
Pain-killing drugs	Treat diseases, relieve pain	Addiction through abuse
Computers	Increased ability to process data	Loss of jobs; health problems from video-display terminals
Space travel	Increased knowledge	High financial cost
Life-sustaining medical devices	Keep people alive	Decisions about when to use or remove them
Automobile	Increased mobility	Increased pollution; deaths and injuries
Chemical fertilizers	Increased agricultural yields	Upset ecology of lakes and streams
Artificial sweeteners	Convenience for diabetics and dieters	Increased risk of cancer

Technological processes and devices should be assessed by their advantages and disadvantages. When a device or process is adopted for use, information on its short-term and long-term effects should be continuously collected and evaluated. This helps us to identify and compare the benefits and possible adverse consequences of the technology for people and the environment, both for present and future generations.

Our society monitors the effects of many technological devices and processes, including medical treatments, food additives, industrial chemicals, and processes for generating electricity. This task is performed by various government agencies and public-interest groups.

Technology and Decision Making

Decisions about the use of technology must be made almost constantly. To make these decisions wisely, both short-term and long-term consequences should be considered. Sometimes the short-term benefits of a technology outweigh its long-term burdens. For example, dentists agree that the benefits of using X rays to find cavities

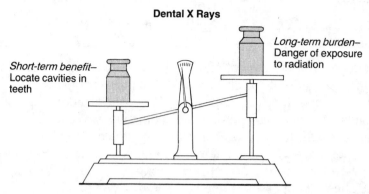

Figure 10-12. Weighing the benefits and burdens of a technology.

in your teeth outweigh the possible long-term dangers of brief exposure to the radiation (Figure 10-12).

In other cases, long-term benefits may outweigh short-term burdens. Wearing a seat belt in a car may be a momentary discomfort. Over time, however, the use of seat belts reduces deaths and injuries from car accidents. Society's consideration of short-term and long-term effects has led to using unleaded gasoline for cars (which is less polluting), and the recycling of cans, bottles, and newspapers (to reduce waste).

The use of a technological device or process may be short-lived or terminated when its disadvantages outweigh its advantages. For instance, in the 1950s a drug was developed to control nausea for pregnant women. When the drug was found to cause an increase in birth defects, its use was banned.

Sometimes a technology can be modified to reduce or eliminate its disadvantages. For example, cars have been modified by the addition of a device called the *catalytic converter* (Figure 10-13). This device greatly reduces the amount of air pollution created by automobiles.

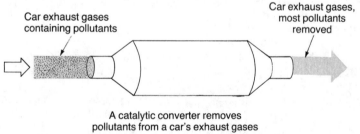

A catalytic converter removes
pollutants from a car's exhaust gases

Figure 10-13. The catalytic converter.

Decisions about complex technological issues usually involve compromises, or *trade-offs*, among alternative courses of action. These trade-offs are often between the benefits and burdens for people and the environment. For instance, electricity generated by burning coal is inexpensive, a benefit to people. But burning coal creates air pollution that leads to acid rain, a burden on the environment. One possible solution is to switch to using oil or natural gas, which burn much cleaner than coal. However, since oil and gas cost much more than coal, this would raise the price of electricity, imposing a burden on the consumer.

An alternative solution is to install devices called *scrubbers* into coal-burning power plants. These devices reduce, but do not eliminate, pollution from burning coal, while increasing electrical costs less than the first solution would. This alternative is a trade-off or compromise between the interests of people and protection of the environment. Such compromises are common in decisions about technology.

PROCESS SKILL: EXPLAINING A RELATIONSHIP

Should government regulate the use of technological devices and processes for the good of society? Some social problems can be solved or lessened by laws that regulate technology. But besides providing benefits, such laws may also burden us in some ways. Some laws may infringe on our right to privacy or freedom of choice. Other laws may be costly to put into action and enforce. Examples of laws that regulate technology are listed in the table below, along with their benefits. On a separate sheet of paper, list some possible burdens of each law.

For instance, the benefit of the seat belt law is a reduction in deaths and injuries from traffic accidents. What burdens does it impose? The law requires car makers to put seat belts in all cars. This raises the price of the car. In many states, the law requires car passengers to wear their seat belts. Some people feel that this interferes with their freedom of choice.

After completing the table, answer the questions that follow.

Laws Regulating Technology

Problem Caused by Technology	Government Regulation	Benefits	Burdens
Automobile accidents	Seat belt law	Fewer deaths and injuries from auto accidents	
Automobile accidents	Lower highway speed limit to 55 m.p.h.	Fewer accidents; better gas mileage	
Air pollution from cars and factories	Smokestack and tailpipe emission controls	Cleaner air	
Air pollution from cigarette smoke	Ban on smoking in public places	Cleaner air	
Bottle and can litter	Bottle deposit bill	Cleaner streets and sidewalks	
Home fire deaths and injuries	Mandatory fire alarms	Fewer deaths and injuries from fires	

1. Which of the following is a burden of the bottle deposit bill?
   (1) less roadway litter
   (2) promotes recycling
   (3) inconvenience of returning bottles to collect refund
   (4) decreases health hazards to wildlife

2. Which of the following is a burden caused by the lower speed limit of 55 miles per hour?
   (1) expense of providing fire alarms
   (2) discomfort of wearing seat belts
   (3) lower public health costs
   (4) longer travel times

3. The following chart shows three more examples of laws regulating technology. Which law has its benefit and burden given in the wrong order?

Problem	Law	Benefit	Burden
Water pollution from industries	Effluent controls law	Cleaner water	Higher industrial costs
Head injuries from motorcycle accidents	Mandatory motorcycle helmet law	Reduces freedom of choice; uncomfortable	Fewer head injuries in motorcycle accidents
Crimes committed with easily obtained hand-guns	Handgun control law	Less crime; harder for criminals to get guns	Harder for people to get guns for protection

(1) effluent controls law
(2) mandatory motorcycle helmet law
(3) handgun control law
(4) they are all in the correct order

EXERCISE 3

1. Decisions about technology should be made by considering
   (1) short-term consequences only
   (2) long-term consequences only
   (3) both short-term and long-term consequences
   (4) the scientific principles on which the technology is based

2. The development of radio and television opened up many new career options for people. This is an example of
   (1) technology developed in response to society's needs
   (2) people's attitudes affecting the acceptance and use of new technology
   (3) increased choices brought about by technology
   (4) technology affecting the global environment

3. Which answer best fills in the blank space in the chart?

Technological Device or Process	Benefit	Burden
Nuclear power plant	Provides efficient, inexpensive energy	

(1) reduces dependence on fossil fuels   (3) creates radioactive waste
(2) provides new jobs   (4) radioactive substances produce heat

4. Which answer lists both a benefit and a burden of television?
   (1) provides entertainment, and provides up-to-date news
   (2) provides entertainment, and causes higher electric bills
   (3) causes eye strain, and discourages reading books
   (4) provides useful educational tool, and provides entertainment

5. Most oil tankers have single hulls that break open fairly easily if the ship runs aground, causing oil spills. However, some tanker ships are now being built with double hulls that are more resistant to breaking. This best illustrates that
   (1) technology has increased our choices in life
   (2) a technology may be modified to reduce or eliminate its disadvantages
   (3) government should constantly monitor technology to determine possible adverse consequences
   (4) decisions about technology often involve trade-offs between benefits and burdens

6. The artificial sweetener saccharin can be safely used by people with diabetes, who cannot eat sugar. It also has fewer calories than sugar and causes less tooth decay. However, experiments have shown that using saccharin increases the risk of getting cancer. Saccharin provides
   (1) a benefit only   (3) both a benefit and a burden
   (2) a burden only   (4) neither a benefit nor a burden

7. In 1984, the bottle deposit bill was passed. This law requires people to pay an extra five cents for every bottle or can of beverage. The deposit is refunded when the empty bottles and cans are returned. The graph indicates that the bottle deposit bill has

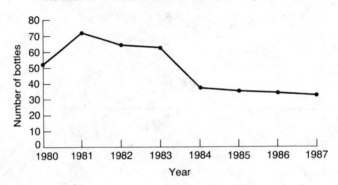

Number of Bottles Collected on a Roadside, 1980–1987

   (1) increased roadside litter
   (2) decreased roadside litter
   (3) had no effect on roadside litter
   (4) its effect cannot be determined from the graph

8. Big cars get poor gas mileage, but they are safer than small cars in the event of a traffic accident. Small cars get very good gas mileage, but offer little protection in an accident. For these reasons, many people buy medium-sized cars, which get fairly good gas mileage and offer some protection in accidents. This is an example of
   (1) a burden of technology on the environment
   (2) monitoring short-term and long-term effects of technology
   (3) a decision about technology that involves a trade-off between benefits and burdens
   (4) modifying technology to reduce or eliminate its drawbacks

9. Although the pesticide DDT was effective in killing insects that damage crops, its use was banned when it was found to be harmful to humans and wildlife. This shows that
   (1) technological products have only disadvantages
   (2) technological products have increased our choices
   (3) use of a technological product may be terminated if its disadvantages outweigh its advantages
   (4) use of a technological product may depend on people's attitudes

Practice Test

1. Which life process releases energy from food?
   (1) respiration   (3) reproduction
   (2) excretion   (4) circulation

2. Many animals in the Arctic have fur that is brown during spring, summer, and fall but turns white in winter. This color change is one way that these animals are
   (1) protected from diseases
   (2) able to attract mates
   (3) adapted to their environment
   (4) able to conserve energy

3. The diagram below shows that
   (1) living things come from other living things
   (2) living things respond to their environment
   (3) living things exchange materials with their environment
   (4) all living things need water to survive

Leaves turn toward light

4. Which organism in the food web shown below is a producer?
   (1) bacteria   (3) grass
   (2) owl   (4) rabbit

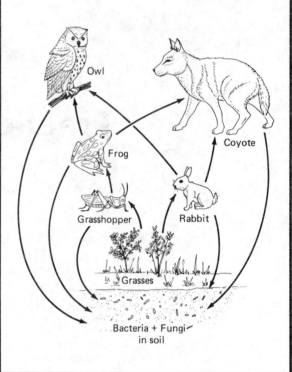

Owl

Frog

Coyote

Grasshopper   Rabbit

Grasses

Bacteria + Fungi in soil

5. A groundhog remains asleep in its underground burrow through the cold winter months. This is an example of
   (1) migration   (3) dormancy
   (2) extinction   (4) hibernation

6. Which member of a food web returns nutrients to the environment?
   (1) producer   (3) decomposer
   (2) consumer   (4) predator

7. Organisms that live in the environment shown in the diagram below have special adaptations to
  (1) gather sunlight
  (2) conserve water
  (3) protect against cold
  (4) breathe underwater

8. The bar graph shows results of a two-week experiment on plant growth and watering. In this experiment, one group of plants (group *A*) was watered four times a week, while a second group of plants (group *B*) was watered two times a week. The graph indicates that, on average, the plants in group *A* grew about
  (1) four times as much as the plants in group *B*
  (2) the same amount as the plants in group *B*
  (3) half as much as the plants in group *B*
  (4) twice as much as the plants in group *B*

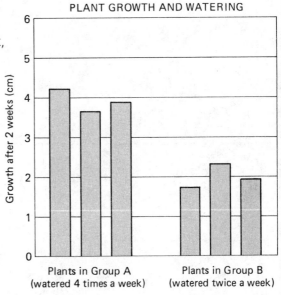

PLANT GROWTH AND WATERING

Growth after 2 weeks (cm)

Plants in Group A
(watered 4 times a week)

Plants in Group B
(watered twice a week)

9. What is the length, in centimeters, of the flatworm shown in the diagram?
  (1) 0.25 cm
  (2) 1.5 cm
  (3) 2.5 cm
  (4) 3.5 cm

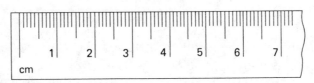

cm

10. A type of muscle that is found only in the heart is called
(1) smooth muscle
(2) involuntary muscle
(3) cardiac muscle
(4) voluntary muscle

11. The esophagus, stomach, and intestines are parts of the
(1) respiratory system
(2) digestive system
(3) circulatory system
(4) excretory system

12. The model shown in the diagram represents which organ system?
(1) digestive system
(2) urinary system
(3) respiratory system
(4) circulatory system

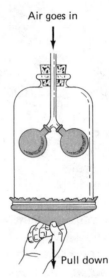
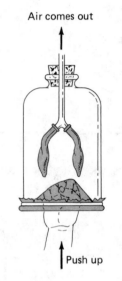

Air goes in

Air comes out

Pull down

Push up

13. The diagram below represents the human circulatory system.

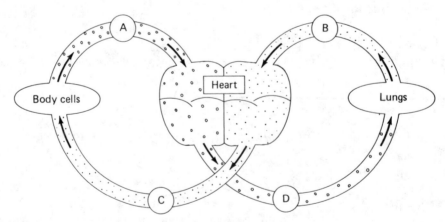

 Blood high in oxygen   Blood high in carbon dioxide

A

B

Heart

Body cells

Lungs

C

D

CIRCULATORY SYSTEM

Compared to the blood in blood vessel *C*, the blood in blood vessel *A* contains more
(1) oxygen   (2) hormones   (3) dissolved nutrients   (4) carbon dioxide

14. In the human male, sperm cells are produced in the
(1) kidneys   (3) ovaries
(2) testes   (4) bladder

15. Materials such as nutrients and wastes are exchanged between the blood and the body's cells at the
(1) capillaries   (3) veins
(2) arteries   (4) air sacs

16. The basic unit of all living things is
(1) the molecule   (3) the nucleus
(2) the organelle   (4) the cell

17. An important difference between sexual and asexual reproduction is that sexual reproduction
(1) involves only one parent
(2) produces offspring identical to the parent
(3) involves only cell division
(4) leads to variation in the next generation

18. Infectious diseases are caused by
(1) harmful microorganisms
(2) toxic chemicals in the environment
(3) a diet lacking essential vitamins and minerals
(4) inherited genetic defects

19. The diagram below compares the structures in a typical plant cell and a typical animal cell. Which structure is found in plant cells but not in animal cells?

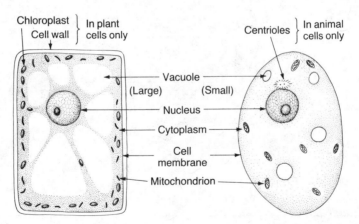

Typical plant cell   Typical animal cell

(1) cell wall
(2) centrioles
(3) cell membrane
(4) mitochondrion

20. The bar graph below shows the number of cases of Lyme disease reported in the United States from 1983 to 1989. In which year was the reported number of cases less than it was the year before?

(1) 1984
(2) 1985
(3) 1986
(4) 1987

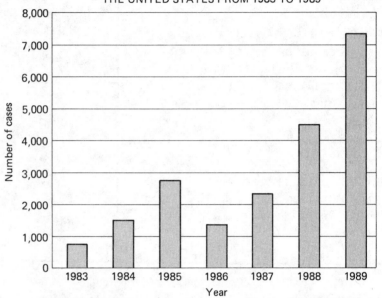

REPORTED CASES OF LYME DISEASE IN THE UNITED STATES FROM 1983 TO 1989

21. The diagram shows undisturbed layers of sedimentary rock. The oldest rocks are those in

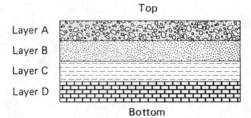

(1) layer *A*
(2) layer *B*
(3) layer *D*
(4) cannot be determined

22. What process is responsible for moving rock material from the continents into the ocean basins?

(1) volcanism   (3) faulting
(2) evaporation   (4) erosion

23. Fossils are found mainly in

(1) igneous rocks
(2) metamorphic rocks
(3) sedimentary rocks
(4) volcanic rocks

24. The diagram at the right shows a mineral with a tendency to break along smooth surfaces in three directions, all at right angles. If the mineral is struck with a hammer, which drawing below shows how the broken pieces will most likely look?

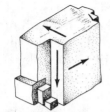

Mineral breaks in 3 directions, at right angles

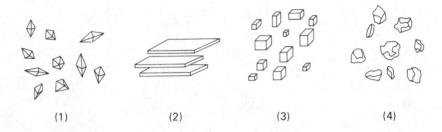

(1)   (2)   (3)   (4)

25. Scientists have found that the ocean floor
(1) has none of the landforms found on the continents
(2) has mountains, valleys, plains, and plateaus
(3) is a flat, featureless plain
(4) is too deep for its features to be studied

26. When two of Earth's crustal plates collide, what surface feature is most likely to be formed?
(1) a new ocean basin
(2) a mountain range
(3) a deep canyon
(4) a continental shelf

27. Air masses and weather fronts usually move across the United States from
(1) east to west   (3) west to east
(2) north to south (4) south to north

28. What temperature is indicated by the thermometer in the illustration below?
(1) 20°C
(2) 15°C
(3) 25°C
(4) 250°C

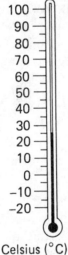

Celsius (°C)

29. The average weather conditions in an area over many years determine that area's
(1) pollution
(2) climate
(3) prevailing winds
(4) earthquake potential

30. The diagram shows a person's exhaled breath turning to mist during cold weather. The process responsible for this is called
(1) evaporation
(2) precipitation
(3) condensation
(4) vaporization

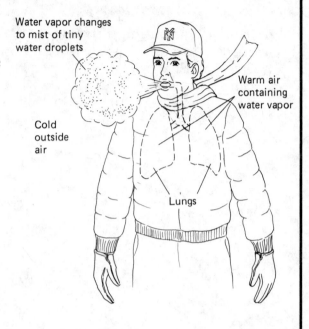

31. The diagram shows a cool air mass pushing into a warmer air mass. The line where the two air masses meet is called a
(1) front
(2) high-pressure system
(3) fault
(4) contour line

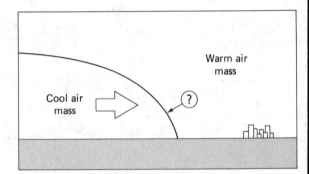

32. The table below shows the daily high and low temperatures for one school week during September in Albany, New York. The *average* temperature (the point halfway between the high and low temperatures) for Wednesday was
(1) 65°F
(2) 120°F
(3) 60°F
(4) 10°F

Day	High Temperature	Low Temperature
Monday	72°F	54°F
Tuesday	70°F	58°F
Wednesday	65°F	55°F
Thursday	68°F	57°F
Friday	69°F	58°F

33. In the water cycle diagram below, the letter *C* represents the process called
  (1) evaporation
  (2) condensation
  (3) radiation
  (4) precipitation

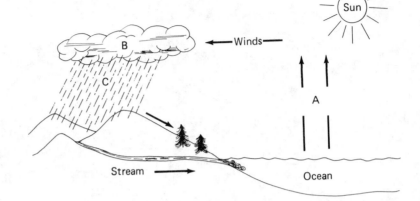

34. The diagram shows the rotating Earth, viewed from above the North Pole (NP). At which point will nightfall take place next?
  (1) *A*
  (2) *B*
  (3) *C*
  (4) *D*

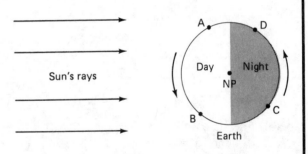

35. The changing seasons on Earth are caused by
  (1) Earth's changing distance from the sun as Earth revolves in its orbit
  (2) the sun changing its heat output during the year
  (3) the drifting of Earth's continents
  (4) the tilt of Earth's axis and Earth's revolution around the sun

36. In New York State, the day with the fewest hours of daylight occurs in the month of
  (1) December
  (2) March
  (3) June
  (4) September

37. The moon revolves around Earth from one full moon to the next in about a
  (1) day
  (2) week
  (3) month
  (4) year

38. The diagram below shows Earth at four positions in its orbit around the sun. When Earth is at position *C*, the season in the northern hemisphere is
(1) summer
(2) winter
(3) fall
(4) spring

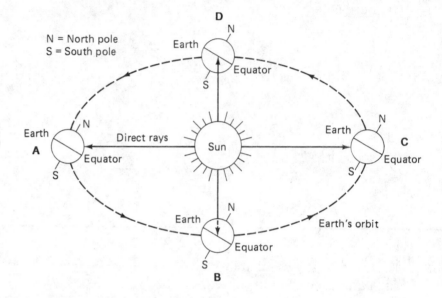

39. The data table below gives the orbital speed and distance from the sun of several planets.

Planet	Orbital Speed (km/s)*	Distance from Sun (km)
Venus	34.8	108,200,000
Earth	29.6	149,600,000
Mars	23.9	227,900,000
Jupiter	12.9	778,000,000
Saturn	9.6	1,427,000,000

*km/s = kilometers per second

The data seem to indicate that

(1) the farther a planet is from the sun, the faster its orbital speed
(2) the farther a planet is from the sun, the slower its orbital speed
(3) the nearer a planet is to the sun, the slower its orbital speed
(4) there is no relationship between a planet's orbital speed and its distance from the sun

40. A streak of light in the night sky that people sometimes call a "shooting star" is actually caused by
(1) a bolt of lightning
(2) an exploding star
(3) a rock fragment entering Earth's atmosphere
(4) a comet orbiting the sun

41. The wheels of a skateboard contain ball bearings in order to reduce
(1) weight   (3) friction
(2) gravity   (4) noise

42. You can see yourself in a mirror because the mirror's surface
(1) reflects light   (3) transmits light
(2) absorbs light   (4) radiates light

43. Which material is the best conductor of electricity?
(1) soil   (3) plastic
(2) copper   (4) glass

44. The handle of a frying pan is often made of plastic rather than metal, because metal
(1) expands when heated
(2) conducts heat well
(3) does not conduct heat well
(4) contracts when heated

45. Compared to the amount of work put into a machine, the amount of work put out by a machine is
(1) always greater
(2) always less
(3) always the same
(4) sometimes greater, sometimes less

46. The screwdriver being used to remove the lid from a can of paint in the diagram below is acting as a
(1) pulley
(2) wedge
(3) inclined plane
(4) lever

47. Heat energy is always transferred
(1) from warmer areas to cooler areas
(2) from cooler areas to warmer areas
(3) from higher regions to lower regions
(4) by materials called insulators

48. The diagram below illustrates that
(1) light travels faster than sound
(2) light travels in straight paths
(3) light can be absorbed, transmitted, or reflected when it strikes an object
(4) light can travel through a vacuum

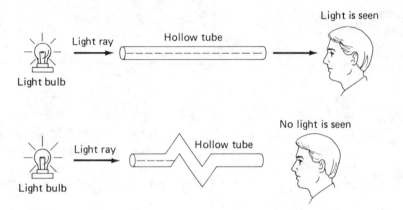

49. The diagrams below show four samples of salt. Each sample contains the same total amount of salt (4 grams). Which sample will dissolve fastest when placed into water?

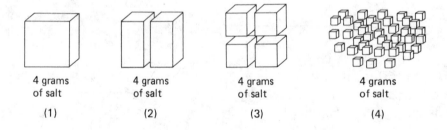

50. Which of the following is *not* an example of matter?
(1) a book
(2) orange juice
(3) a sound
(4) steam from boiling water

51. Chopping a carrot into small slices is an example of a
(1) physical change
(2) chemical change
(3) change in phase
(4) chemical reaction

52. As the temperature of a sample of water increases, the amount of sugar that can be dissolved in the water also increases. Which graph correctly shows this relationship?

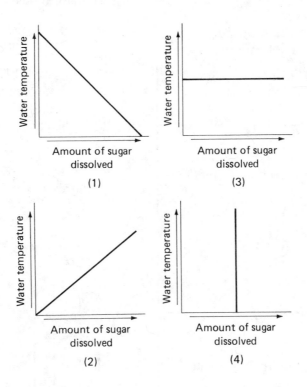

53. The table below shows some "English" units of measurement (which are familiar to most Americans) and their equivalents in the metric system. Using the table and some simple math, you can tell that 3 kilograms is equivalent to
  (1) 1 pound
  (2) 2.2 pounds
  (3) 454 pounds
  (4) 6.6 pounds

English Unit	Metric Equivalent
1 inch	2.54 centimeters
39.4 inches	1 meter
1 pound	454 grams
2.2 pounds	1 kilogram
1.06 quarts	1 liter

54. "Warning: contents under pressure. Do not puncture or incinerate." This warning would most likely be found on a
(1) bottle of aspirin
(2) can of crystal drain opener
(3) bottle of household ammonia
(4) can of hair spray

55. When a substance undergoes melting, it changes from
(1) a liquid to a gas
(2) a gas to a liquid
(3) a solid to a liquid
(4) a liquid to a solid

56. What property of the rock is being tested in the diagram?

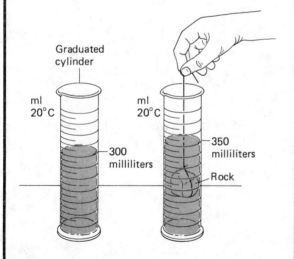

(1) volume   (3) mass
(2) shape   (4) hardness

57. Which of the following is *not* a fossil fuel?
(1) coal   (3) uranium
(2) natural gas   (4) oil

58. The bar graph below compares the amount of foreign oil used daily in the United States to the amount of domestically produced oil used daily.

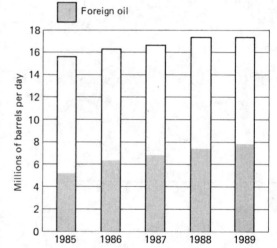

In 1989, about how much foreign oil was used daily in the U.S.?
(1) 6 million barrels
(2) 10 million barrels
(3) 8 million barrels
(4) 17 million barrels

59. Which energy source is a renewable resource?
(1) coal   (3) moving water
(2) uranium   (4) oil

60. Two major problems associated with the use of nuclear energy are thermal pollution and
(1) environmental damage from strip mining
(2) acid rain caused by air pollution
(3) disposal of radioactive wastes
(4) black lung disease

Questions 61–63 refer to the two pie graphs below, which show percentages of energy sources in world energy consumption for the years 1939 and 1989.

WORLD ENERGY CONSUMPTION

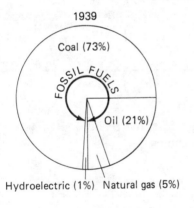

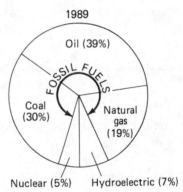

61. Between 1939 and 1989, the percentage of the world's energy supplied by fossil fuels
(1) increased
(2) decreased
(3) stayed the same

62. In 1989, what percentage of world energy consumption was supplied by energy sources other than the fossil fuels?
(1) 12%   (3) 7%
(2) 5%   (4) 88%

63. What energy source was available in 1989 that was not available in 1939?
(1) hydroelectric
(2) natural gas
(3) nuclear
(4) oil

64. What energy source is being used by the devices shown in the diagram below?

(1) solar energy

(2) nuclear energy

(3) wind energy

(4) hydroelectric energy

65. The table below lists scientific principles and some technological devices or processes based on those principles.

	Scientific Principle	Technological Device or Process
Line A	Sunlight contains energy	Solar cells, solar heating systems
Line B	Telescopes and microscopes	Lenses bend light rays
Line C	Splitting atoms releases energy	Nuclear power plants, nuclear weapons
Line D	Cold temperatures reduce growth of microorganisms	Refrigerators and freezers

Which line of the table has the scientific principle and the technological device listed in the wrong columns?

(1) line A

(2) line B

(3) line C

(4) line D

66. When seatbelts were first introduced in cars, they consisted of a single strap across the waist. However, this strap allowed a person's upper body to snap forward during an accident, sometimes causing head injuries. To remedy this, a strap was added that crosses the chest and goes over one shoulder. The shoulder strap prevents the upper body from flying forward in an accident. This shows that

(1) science and technology help each other advance
(2) people's attitudes can affect acceptance and use of new technology
(3) a technological device may be modified to reduce or eliminate its disadvantages
(4) use of a technological device may be short-lived if its disadvantages outweigh its advantages

67. Global warming caused by air pollution resulting from the burning of fossil fuels is an example of

(1) an impact of technology on just one country
(2) science and technology helping each other advance
(3) a benefit and a burden of technology
(4) an international impact of technology

68. Asking questions and seeking their answers in order to gain knowledge about the natural world is known as

(1) technology
(2) science
(3) industry
(4) society

69. A jet airplane is an example of

(1) a simple technological device
(2) a complex technological device made up of a number of simpler devices working together
(3) several simple devices that are unrelated to each other
(4) a technological device that is becoming obsolete

70. The data table below shows high, low, and average monthly temperatures on the South Island of New Zealand for half a year.

Month	Temperatures (°F)		
	High	Low	Average
Jan.	85	41	63
Feb.	83	41	62
Mar.	81	37	59
Apr.	75	33	54
May	69	29	49
June	62	26	44

Construct a graph of the *average* monthly temperatures given in the table by copying the numbered axes shown in the sample below onto a separate sheet of graph paper. Plot the data for the average monthly temperatures onto the graph and connect the points with a line.

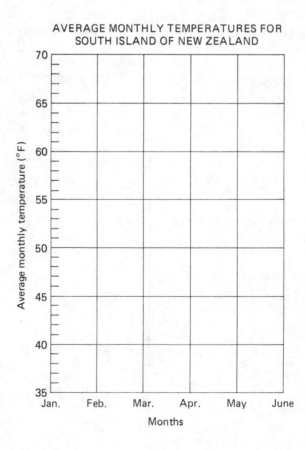

AVERAGE MONTHLY TEMPERATURES FOR
SOUTH ISLAND OF NEW ZEALAND

Glossary

Acid rain: Rain that has been made more acidic than normal by pollutants in the atmosphere.

Adaptation: A characteristic that helps an organism survive in its habitat.

Air mass: A large body of air that has roughly uniform temperature and humidity throughout.

Air pressure: The force with which air presses down on Earth's surface.

Air temperature: A measurement of the amount of heat energy in the atmosphere.

Altitude: The height above sea level of a place.

Arteries: Blood vessels that carry blood away from the heart.

Asexual reproduction: Reproduction that involves only one parent, producing offspring that are genetically identical with the parent.

Atom: The smallest particle of an element that has the properties of that element.

Axis of rotation: An imaginary line through an object, around which the object spins.

Bedrock: The unbroken, solid rock portion of Earth's crust.

Benefit: An advantage to a person or society of a technological device or process.

Blood: A liquid tissue that contains red and white blood cells, and platelets, and also carries dissolved gases, nutrients, hormones, and wastes.

Blood vessels: Tubes through which the blood flows.

Boiling: The rapid change in phase from liquid to gas, during which bubbles of gas form within the liquid.

Boiling point: The temperature at which a substance changes rapidly from a liquid to a gas.

Brain: The organ, located within the skull, that controls thinking and body activities.

Bronchi: The two tubes that branch off from the lower end of the trachea, connecting it to the lungs.

Burden: A disadvantage to a person or society of a technological device or process.

Calorie: A unit used to measure and compare the amount of heat energy contained in substances such as foods and fuels.

Canning: The process of preserving food by sealing it in airtight containers, which are then sterilized by heating.

Capillaries: Tiny blood vessels, connecting arteries to veins, through which materials are exchanged between the blood and the body's cells.

Cartilage: A flexible tissue that acts as a cushion between bones and provides flexibility at the ends of bones.

Cell: The basic unit of all living things.

Cell division: The process by which cells reproduce, wherein a parent cell splits into two new daughter cells.

Cell membrane: The outer covering, or "skin," of a cell, which controls the flow of materials into and out of the cell.

Cellular respiration: A life process that occurs in all cells, in which nutrients from digested food are combined with oxygen to release energy and produce the wastes carbon dioxide and water.

Chemical bond: The link that joins one atom to another in a molecule.

Chemical change: A change that results in the formation of one or more new substances; a chemical reaction.

Chemical property: A characteristic that a substance displays when it undergoes a change to a new substance or substances.

Circuit breaker: A device that prevents overloading of an electric circuit by interrupting the flow of electricity when it reaches a dangerous level.

Climate: The general character of the weather in an area over many years.

Cloud: A mass of tiny water droplets or ice crystals suspended high in the atmosphere.

Coal: A black rock formed from the remains of ancient swamp plants. Coal is a fossil fuel.

Cold front: The boundary formed when a cool air mass pushes into and under a warm air mass.

Community: All the different organisms that live within a habitat.

Condensation: The changing of water vapor into droplets of liquid water; more generally, the change in phase from gas to liquid.

Conductor: A material through which electricity can flow.

Conservation: The saving of natural resources through wise use.

Consumer: An organism that obtains nutrients by eating other organisms.

Crust: The outermost rock layer of Earth, which contains all of Earth's surface features.

Cytoplasm: The watery substance that fills the cell, where most life processes occur.

Decomposer: An organism that breaks down the remains and wastes of other living things.

Disinfection: The destruction of all or most of the harmful microorganisms in an area by using chemicals called germicides.

Dormancy: A state in which an organism is inactive while it awaits more favorable conditions in its environment.

Earthquake: A shaking or vibrating of Earth's crust, usually caused by the sudden movement of rocks sliding along a fault.

Ecological succession: The natural process by which one community of living things is replaced by another community, until a stable climax community appears.

Ecosystem: The living members of a community, plus the nonliving elements of their environment.

Electric circuit: A complete path for the flow of electricity.

Electricity: A form of energy produced by the flow of electrons from one point to another point.

Electromagnetic waves: Energy waves that travel at the speed of light and can move through a vacuum; they include radio waves, microwaves, infrared waves, visible light, ultraviolet waves, X rays, and gamma rays.

Element: One of the 109 known basic substances that form the building blocks of matter.

Energy: The ability to do work.

Environment: The surroundings in which an organism lives, including both living and nonliving things.

Erosion: The process whereby rock material at Earth's surface is removed and carried away.

Evaporation: The changing of liquid water into water vapor (gaseous water); more generally, the change in phase from liquid to gas.

Faulting: The process in which internal forces cause Earth's crust to break and slide along a fracture called a fault.

Fertilization: The joining of an egg cell and a sperm cell, during sexual reproduction, to begin the development of a new individual.

Flammable: Capable of catching fire and burning easily.

Folding: The process whereby rock layers in Earth's crust are squeezed into wavelike patterns called folds.

Food chain: A sequence of organisms through which nutrients are passed along in an ecosystem.

Food web: A number of interconnected food chains.

Fossil: The remains or traces of an ancient organism.

Fossil fuel: A fuel that was formed from the remains of ancient plants or animals; examples include oil, coal, and natural gas.

Freezing: (1) The change in phase from liquid to solid. (2) The storing of food at temperatures below 0 °C (32 °F), to slow the growth of microorganisms that can spoil food.

Freezing point: The temperature at which a substance changes from a liquid to a solid.

Front: The boundary between two different air masses.

Full moon: The phase of the moon that occurs when Earth is between the sun and the moon, so that all of the moon's lighted side can be seen from Earth.

Fuse: A device, used in an electric circuit, containing a thin metal strip that melts to interrupt the flow of electricity when the circuit becomes overheated.

Gland: An organ that makes and secretes (releases) chemicals called hormones.

Greenhouse effect: The trapping of heat in Earth's atmosphere by carbon dioxide, leading to global warming.

Grounding: A safety feature of an electrical device or circuit in which an extra wire attached to the device or circuit conducts any excess charge to the earth, where it is absorbed.

Habitat: The particular environment in which an organism lives.

Heart: An organ, made mostly of muscle, that contracts (beats) regularly to pump blood throughout the body.

Heat energy: The energy of motion of the vibrating particles that make up matter.

Hibernation: A sleeplike state of reduced body activity that some animals enter to survive the winter.

High-pressure system: A large area where air is sinking, causing high surface air pressure; also called a high.

Hormone: A chemical "messenger" secreted by a gland into the bloodstream, which carries the hormone to an organ that responds in some way.

Humidity: The amount of moisture (water vapor) present in the atmosphere.

Hurricane: A huge, rotating storm that forms over the ocean in the tropics, with winds of over 75 miles per hour and heavy rains.

Hydroelectric energy: Electricity produced by using the energy of flowing water to turn the turbines of a generator.

Igneous rock: A rock formed by the cooling and hardening of hot, liquid rock material.

Inclined plane: A simple machine that consists of a flat surface with one end higher than the other, such as a loading ramp.

Infectious disease: A disease caused by microorganisms that can be transmitted from one individual to another.

Insulation: Material used to reduce or slow the flow of heat from one area to another.

Insulator: A material through which electricity cannot flow.

Involuntary muscles: Muscles that we do not consciously control.

Joint: A place where one bone is connected to another bone.

Kidneys: A pair of organs that filter wastes from the blood and help control the water and mineral balance of the body.

Kinetic energy: Energy that an object has because of its motion.

Latitude: Distance from the equator, measured in degrees.

Lens: A piece of transparent glass or plastic with curved surfaces that bend light rays to form an image.

Lever: A simple machine consisting of a bar or rod that can turn around a point called the fulcrum.

Life cycle: The changes an organism undergoes as it develops and produces offspring.

Light: A visible form of energy.

Light-year: The distance that light travels in one year, about 9.46 trillion kilometers.

Liver: An organ that produces urea from excess amino acids, removes harmful substances from the blood, and secretes bile, a digestive juice.

Locomotion: The movement of the body from place to place.

Low-pressure system: A large area where air is rising, causing low surface air pressure; also called a low.

Lungs: A pair of organs, located in the chest, that contain millions of tiny air sacs, in which the exchange of respiratory gases between the blood and the environment takes place.

Lymph: A fluid that bathes all body cells and acts as a go-between in the exchange of materials between the blood and the cells.

Lymph vessels: Tubes in which waste-laden lymph is collected and returned to the bloodstream.

Machine: A device that transfers mechanical energy from one object to another object.

Mammary glands: The female breasts, which produce milk to nourish the newborn offspring.

Mass: The amount of matter in an object.

Matter: Anything that has mass and takes up space.

Melting: The change in phase from solid to liquid.

Melting point: The temperature at which a substance changes from a solid to a liquid.

Metamorphic rock: A rock produced when existing igneous or sedimentary rock undergoes a change in form caused by great heat, pressure, or both.

Meteor: A rock fragment traveling through space that enters Earth's atmosphere and burns up, producing a bright streak of light.

Microorganism: An organism that is very small, usually too small to be seen with the unaided eye.

Migration: Moving from one environment to another, where conditions are more favorable.

Mineral: A naturally occurring, solid inorganic substance with characteristic physical and chemical properties.

Mountain: A feature on Earth's surface that rises relatively high above the surrounding landscape.

Muscles: Masses of tissue that contract to move bones or organs.

Natural gas: A gaseous fossil fuel found trapped deep underground, often with oil deposits.

Nerves: Thin strands of tissue, composed of neurons, that carry impulses throughout the body.

Neurons: Cells that make up the nervous system, which receive and transmit information in the form of impulses.

New moon: The phase of the moon that occurs when the moon is between Earth and the sun, so that the moon cannot be seen from Earth.

Noninfectious disease: A disease that cannot be transmitted from one individual to another.

Nonrenewable resource: A resource that is not replenished by nature within the time span of human history.

Nuclear energy: The energy stored within the nucleus of an atom, used by nuclear power plants to produce electricity.

Nuclear waste: The poisonous, radioactive remains of the materials used to fuel nuclear power plants.

Nucleus: (1) The structure within the cell that controls cell activities and contains genetic material. (2) The center of an atom.

Nutrients: Food substances that supply an organism with energy and with materials for growth and repair.

Oil: A thick, black, liquid fossil fuel, found trapped underground; also called petroleum.

Orbit: The path of an object in space that is revolving around another object.

Organ: A group of tissues that act together to perform a function.

Organism: A living thing.

Organ system: A group of organs that act together to carry out a life process.

Ovaries: The female reproductive organs that produce egg cells.

Oviducts: Tubes that connect the ovaries to the uterus.

Pasteurization: The process of heating milk, and some other foods, to kill bacteria that cause spoilage and disease.

Phases: (1) The changing apparent shape of the moon, as seen from Earth. (2) The three forms, or states, of matter—solid, liquid, and gas.

Photosynthesis: The process by which green plants produce food, using sunlight, carbon dioxide, and water; oxygen is given off as a by-product.

Physical change: A change in the appearance of a substance that does not alter the chemical makeup of the substance.

Physical property: A characteristic of a substance that can be determined without changing the identity of the substance.

Plain: A broad, flat landscape region at a low elevation, usually made of layered sedimentary rocks.

Plateau: A large area of Earth's surface made up of horizontally layered rocks, found at a relatively high elevation.

Plate tectonics: The theory that Earth's crust is broken up into a number of large pieces, or plates, that move and interact, producing many of Earth's surface features.

Pollutants: Harmful substances that contaminate the environment, often produced by human activities.

Potential energy: Stored energy that an object has because of its position or chemical makeup.

Precipitation: Water, in the form of rain, snow, sleet, or hail, falling from clouds in the sky.

Prevailing winds: The winds that commonly blow in the same direction at a given latitude.

Producer: An organism that makes its own food. Most producers are green plants.

Refrigeration: The storing of food at cold temperatures to slow the growth of harmful bacteria.

Renewable resource: A resource that is replenished by nature within a relatively short time span.

Reproduction: The life process by which organisms produce new individuals, or offspring.

Respiration: (1) The process of taking in oxygen from the environment and releasing carbon dioxide and water vapor. (2) See also **cellular respiration**.

Response: The reaction of a living thing to a change in its environment.

Revolution: The movement of an object in space around another object, such as the revolution of the moon around Earth.

Rock: A natural, stony material composed of one or more minerals.

Rotation: The spinning of an object around its axis.

Science: The study of the natural world.

Sedimentary rock: A rock formed from layers of particles, called sediments, that are cemented together under pressure.

Sense organs: Organs that receive information from the environment. The sense organs include the eyes, ears, nose, tongue, and skin.

Sexual reproduction: Reproduction that involves two parents, producing offspring that are not identical with either parent.

Skin: The organ that covers and protects the body, and excretes wastes by perspiring.

Smog: A haze in the atmosphere produced by the reaction of sunlight with pollutants from cars and factories.

Soil: A mixture of small rock fragments and decayed organic material that covers much of Earth's land surface.

Solar energy: Energy from the sun.

Solar system: The sun and all the objects that revolve around it, including the planets and their moons, asteroids, comets, and meteors.

Sound: A form of energy produced by a vibrating object.

Sound waves: Alternating layers of compressed and expanded air particles that spread out in all directions from a vibrating object.

Species: A group of organisms of the same kind.

Sperm ducts: Tubes through which sperm cells pass upon leaving the testes.

Spinal cord: The thick cord of nerve tissue that extends from the brain down through the spinal column.

Sterilization: The killing of all microorganisms in an area, usually by heating.

Stimulus: A change in the environment that causes an organism to react in some way.

Storm: A natural disturbance in the atmosphere that involves low air pressure, clouds, precipitation, and strong winds.

System: A group of related elements or parts that work together for a common purpose.

Technology: The application of scientific knowledge and other resources to develop new products and processes.

Testes: The male reproductive organs that produce sperm cells.

Thermal pollution: An increase in the temperature of a body of water, caused by human activities, that may be harmful to living things in that environment.

Thunderstorm: A brief, intense rainstorm that affects a small area and is accompanied by thunder and lightning.

Tides: The rise and fall in the level of the ocean's waters that take place twice each day.

Tissue: A group of similar cells that act together to perform a function.

Tornado: A violent whirling wind, sometimes visible as a funnel-shaped cloud.

Toxic: Poisonous.

Trachea: The tube that connects the nose and mouth to the bronchi, which lead to the lungs; also called the windpipe.

Uranium: A radioactive element found in certain rocks and used as a fuel for nuclear power plants.

Uterus: The organ of the female reproductive system within which an offspring develops; also called the womb.

Vapors: Fumes or gases given off by a substance.

Veins: Blood vessels that return blood to the heart.

Volcano: (1) An opening in Earth's surface through which hot, liquid rock flows from deep underground. (2) A mountain formed by a series of volcanic eruptions.

Volume: The amount of space an object occupies.

Voluntary muscles: Muscles that we consciously control.

Warm front: The boundary formed when a warm air mass slides up and over a cool air mass.

Water cycle: The process in which water moves back and forth between Earth's surface and the atmosphere by means of evaporation, condensation, and precipitation.

Watt: A unit that measures the rate at which electrical energy is used.

Weather: The changing condition of the atmosphere, with respect to heat, cold, sunshine, rain, snow, clouds, and wind.

Weathering: The breaking down of rocks into smaller pieces.

Wind: The movement of air over Earth's surface.

Wind direction: The direction from which the wind is blowing.

Work: The moving of an object over a distance by a force.

Index

Note: "t" and "f" denote table and figure references.